GreenTrading™ creates solutions for reducing pollution and accelerates technology transfer. The GreenTrading™ marketplace encompasses the trading of credits associated with:

- Carbon and greenhouse gas emission reduction;
- Renewable energy; and
- Energy efficiency (Negawatts)

This new markets allow the creation of quantifiable commodities that are fungible across markets and borders. With appropriate public policies and market structures, ***Triple Convergence*** of these markets will create liquidity, facilitate the financing of environmental improvements, and accelerate technological adoption. Portfolios of such commodities are expected to appreciate in value over time.

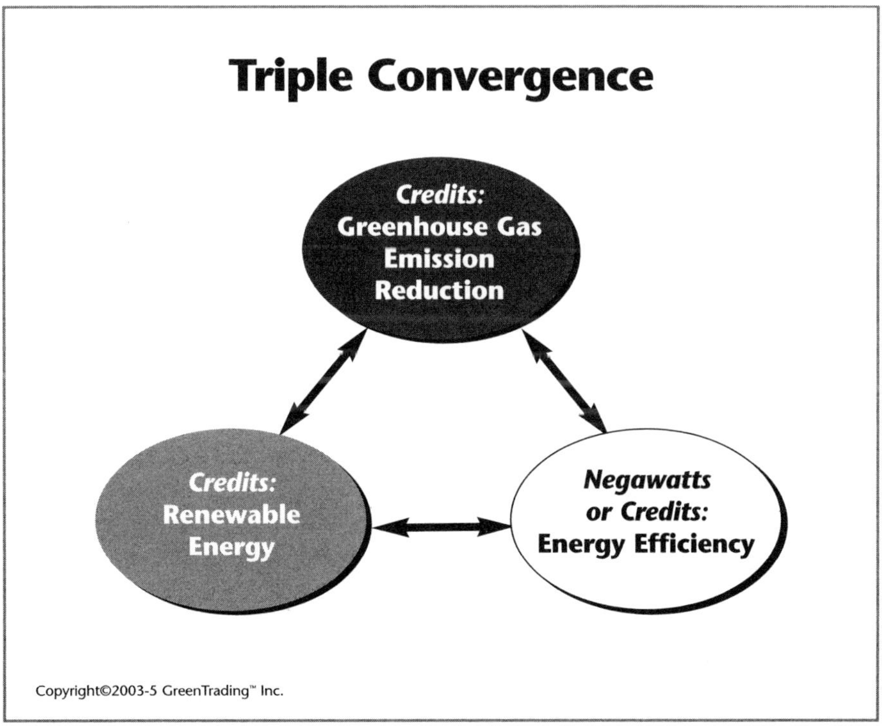

Green Trading Markets: Developing the Second Wave

Peter C. Fusaro and
Marion Yuen

GreenTrading™

ELSEVIER

Amsterdam Boston London New York Oxford Paris San Diego San Francisco Singapore Sydney Tokyo

Elsevier Science Ltd
The Boulevard Langford Lane
Kidlington, Oxford OX5 1GB

Elsevier Science Ltd.
The Boulevard, Langford Lane
Kidlington, Oxford OX5 1GB
United Kingdom

Catalog-in-Publication Data

Peter C. Fusaro and Marion Yuen
Green Trading Markets: Developing the Second Wave
ISBN 008044695-7
1. Investing. 2. Finance. 3. Environment. 4. Risk Management. 5. Energy Trading.

Additional books can be ordered online at http://www.elsevier.com. Consult the Elsevier homepage for full catalogue information on all books, journals and electronic products and services.

Special quantity discounts are available for use in education, corporate training, or promotional programmes. For more information, please email Judy Chappell at J.Chappell@elsevier.co.uk.

Design by S. Kloos Communications Inc.
Cover Image by StockTrek/Getty Images

Printed on acid-free paper in the United States of America

First Edition 2005

This book is dedicated to the great people of New York City.

Contents

Contents *(continued...)*

Foreword

Can a healthy economy, sound energy policy, and clean environment coexist? Can business create environmental opportunities that are profitable? In the post–9/11 United States, will environmental issues gain traction?

The answer to all of the above is a resounding "Yes."

The chapters of this book weave a story of an emerging environmental economy. Business leaders will find ideas to sustain high performance and develop new markets. Government workers will find innovative market-driven policy options stimulating the economy and creating environmental improvements. This integration of finance and the environment is being called the era in which Adam Smith meets Rachel Carson.

A second wave of environmental activity is emerging, where businesses will see strategic opportunities to create profitable growth, derive competitive advantages and better manage risk, while improving the quality of our land, air, and water resources.

Real-life examples abound:

- Large industrial companies are preparing for the day when carbon emissions come with a price tag. British Petroleum, Shell Oil, and DuPont are defining the nature and extent of their carbon footprint, baselining their emissions and preparing to trade carbon emission reduction credits in the open market for a profit.
- Over 20 states have Renewable Energy Portfolio Standards (RPS) mandating that a certain percentage of their energy comes from alternative renewable sources. These programs offer substantial tax breaks, provide rebates for capital expenditures, and create a market for Renewable Energy Credits (RECs) that can be bought and sold.
- Such companies as General Electric, Ford Motor Company, and General Motors are working with their supply chain, exploring ways to lower costs, create operational improvements, and reduce risk by strengthening environmental/energy practices. At the same time, these companies are seizing opportunities to manufacture environmentally sound products ranging from wind turbines to hybrid cars.

A hidden reserve of opportunity exists in environmental programs. The first chapter of this book, written by Peter Fusaro, details the convergence of capital markets and the environment, where improvements are being financed by innovative programs in energy and harnessing the power of market forces, such as green trading. Other chapters describe the political landscape of potential federal "cap and trade" programs for environmental trading, and how state actions are encouraging the emerging market for environmental commodities such as RECs and carbon offsets. Finally, the remaining chapters discuss the impact of Socially Responsible Investing (SRI) practices on business decisions regarding the environment. Over $2.16 trillion in assets worldwide, and one out of every nine investment dollars in the United States, are being managed using SRI strategies. This creates a "fiduciary capitalism" in which shareholders are making investment decisions based on corporate governance issues involving the environment and energy.

As Chairman and Chief Executive Officer of Tetra Tech, I am seeing tangible examples of this new industrial revolution that links environment, energy, and finance together. For example, we are helping a large racetrack evaluate the use of horse manure, straw, and other refuse as a resource for energy generation. This innovative idea lowers power costs, creates reoccurring revenue from RECs, and provides an innovative waste disposal solution in their community by eliminating the need for more landfill space and reducing the environmental impacts of transporting waste product.

In addition, large manufacturing facilities, schools and residences across the county are designing and building solar power systems to save on life cycle energy expenses, reduce dependence on fossil fuels, curb harmful emissions; they also are able to sell any extra power to the grid for a profit. The underlying cash flow from trading RECs is being utilized to help finance such capital improvements. Schools are integrating these solar systems into their curriculum, providing a learning opportunity for their students concerning sustainability, the importance of renewable energy and integration of business with environment.

As an environmental engineering and consulting company, Tetra Tech helps clients deal with the "whole environment" from hazardous waste remediation, water quality, environmental health and safety, risk management, green building design, megawatts programs, and renewable energy development. I very much believe that we are at the forefront of a market-based environmentalism that will create business opportunities at the nexus of economy, energy, environment, and finance.

As Tetra Tech is the first environmental and engineering consulting company to become a member of the Dow Jones Sustainability Index, we are confident the next wave of the environmental economy will be a pillar of growth for us. The following pages provide a roadmap for the reader to better understand these issues and discover their own unique opportunities.

Li-San Hwang
Chairman and CEO
Tetra Tech

November 2004

Meeting the World's Growing Energy Needs

While renewable energy holds great promise, no single solution will be able to meet all of the world's growing requirements for electrical power. Satisfying that demand will require a balanced portfolio of energy options ranging from natural gas, coal, and nuclear to wind, and hydro, and in the future including new technologies such as solar photovoltaics, and hybrid fuel cells.

A common theme runs through all of the disparate energy technologies available today: Produce power efficiently, while having the least possible impact on the environment. That initiative takes on many shapes.

For gas turbines, higher fuel efficiency and advanced emissions control technology are leading to improved output relative to emissions. Forty years ago, simple-cycle gas turbines operated at thermal efficiencies of 28 to 29 percent; today's natural gas-fired, combined-cycle systems can reach 60 percent. This allows tremendous increases in power output at a lower rate of fuel consumption.

For both new gas turbines and for retrofits of older, installed machines, emissions technology continues to evolve and improve. In the 1960s, gas turbine NOx emissions levels of 200 parts per million were common; today's advanced technology can drive gas turbine emissions into single digits.

Coal remains one of the world's most abundant fossil fuels. Developments in selective catalytic reduction, flue gas scrubbers, fluidized bed technology, and particularly coal gasification now make it possible to burn coal with lower emissions than ever before.

Any discussion about meeting the world's energy needs also needs to include nuclear power. The lack of carbon depletion and avoidance of millions of tons of emissions (NOx, SO_2 and CO_2) make nuclear power one of the few clean energy sources capable of generating huge power output.

But key issues must be resolved before nuclear energy will enjoy a resurgence, particularly in the U.S. Acceptable solutions must be found for the transportation and storage of nuclear waste; the industry needs a supportive administration over a long period of time; and the technology's risks and rewards must be thoroughly understood by the investing community.

After years of facing significant economic and technical challenges, renewable energy is making great strides forward. Today, it is being increasingly recognized worldwide as a viable means to reduce the threat of global climate change, encourage development, and create jobs.

Much of the growth in the renewables sector has been driven by the rapidly expanding wind energy industry. As wind turbine technology continues to advance and the installed base of wind turbines grows, the cost of wind-generated electricity is becoming competitive with other energy options—a key factor in the growing acceptance of the technology.

Today's wind technology has grown not only in size, but also has become increasingly sophisticated and highly reliable. An example is GE's 3.6-megawatt wind turbine, the first wind turbine over three megawatts designed specifically for offshore applications. The machine offers patented power electronics and a variable speed rotor for cost-effective, reliable operation.

While wind power has the fastest growth rate, hydropower is the most established renewable energy technology. According to the U.S. Department of Energy (DOE), in 2003, hydropower supplied more than 75 percent of the electricity generated by renewable sources in the U.S., and about 10 percent of the country's total electricity capacity.

Hydropower will continue to play a significant role in the overall energy picture. The DOE forecasts a 56 percent increase in renewable energy use by 2025, with most of the increase coming from new, large-scale hydroelectric plants in developing countries.

Other renewable technologies such as solar, biodiesel, and geothermal are expected to make increasingly significant contributions to the world's energy mix in the years ahead.

Solar, in particular, has strong potential, since solar cells can be placed nearly anywhere. Already a cost-effective source of power in many remote, off-grid locations, solar cells could help meet the power needs of many of the two billion people worldwide who now lack access to modern energy services.

Despite their great potential, the various renewable energy technologies that exist today will not, in and of themselves, come close to generating all of the new power capacity the world will demand in the future. Clearly, they will be a vital part of the equation, but the answer will come from a wide range of energy technologies that will continue to improve and drive environmental preservation and efficiency as the top priorities.

John Rice
President and CEO
GE Energy

December 2004

Developing the Second Wave of Green Trading

By Peter C. Fusaro and Marion Yuen

About two years ago, we began to notice signs of the second wave of Green Trading, also called environmental financial trading. Earlier this year (2005), the Kyoto Protocol and the European Union Emissions Trading Scheme (E.U.-E.T.S.) entered into effect at the international level. In the United States, the Chicago Climate Exchange began its second year of operations and, in late 2004, extended its presence to Europe with the creation of the European Climate Exchange. Sulfur Financial Instruments are now being traded on the Chicago Climate Futures Exchange, and the New York Mercantile Exchange has announced that it will soon launch SO_2 and NOx futures contracts. Sometime in 2005, we are likely to see the enactment of the RGGI (Regional Greenhouse Gas Initiative), which is the first U.S. mandatory cap-and-trade program for carbon dioxide in the northeastern and mid-Atlantic states.

This book, and the series of Annual GreenTrading™ Summits from which it originated, are dedicated to promoting the development of the U.S. Green Trading markets and the convergence of its components—the trading of credits in greenhouse gas (GHG)[1] reduction, renewable energy, and energy efficiency. These component markets and their regional forms will grow and achieve greater liquidity as financial instruments are introduced and accepted, and as the financial products become more fungible with the support of appropriate market structures and regulations. Our vision is the convergence of these markets with the capital markets, allowing them to attain maximum effectiveness and accelerate investment of clean, environmentally benign energy technologies.

Figure 1

The Triple Convergence in Green Trading

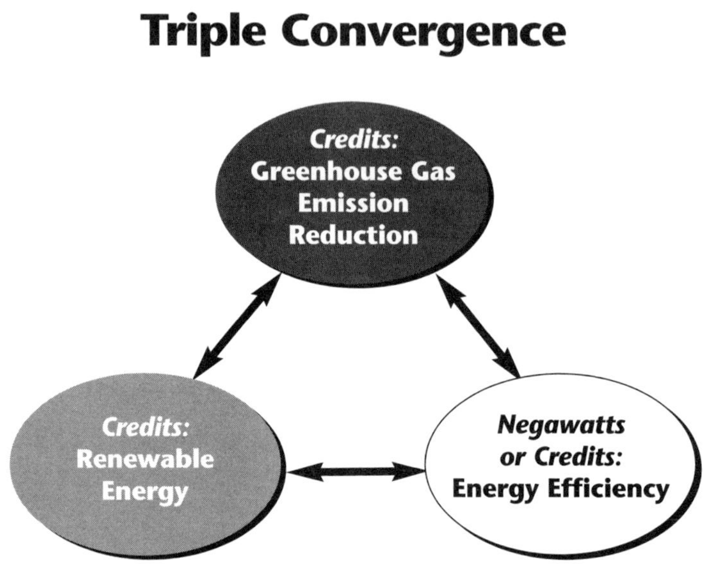

Why Green Trading? The basis of the world economy is undergoing a fundamental shift, comparable to the transitions when coal substituted for wood as the primary fuel, when oil displaced coal, and when homes lighted by candles and whale oil began to use gas and then, electricity. Unlike the previous economies, this new energy order is not simply necessitated by the increasingly difficult and costly access to fuel sources. In addition, it is impacted by the response of the natural environment to the byproducts from the use of fossil fuels (including the buildup of carbon dioxide and other greenhouse gases in the atmosphere) and the unintended results of economic development that were made possible by the availability of oil and coal (such as large-scale land clearance and desertification). The environmental response is global climate change as we have come to know it in the last decades. Unlike previous transitions, the current shift to a carbon-constrained economy is global in scale and affects all regions, albeit unevenly.

The current transition is straining existing economic structures, as evidenced by the high volatility of oil prices as well as the higher prices and volatility of all natural resources. In the new economy, where carbon promises to be the new gold, economic values will be distributed differently from the current fossil fuel-based economy and new cultural values will emerge, giving rise to new economic and political expressions. An example of this cultural shift is the willingness of consumers, over the past 10 years or so, to pay a premium for green power (electricity generated from renewable or nonpolluting sources) and support organized pressure on local and state governments to mandate Renewable Portfolio Standards or adopt renewable set-asides in energy purchase. Increasing numbers of consumers are willing to support something seemingly intangible such as "environmental attributes," based on their personal values.

As with any change, there is the vanguard. In *Carbon Down, Profits Up* (2004), the Climate Trust collected the results of corporations and governmental entities that have seen the new paradigm, taken proactive measures, and begun to profit from their initiatives. Nevertheless, such efforts require money as well as departure from familiar practices and established standards, use of new technologies, and new uses of known technologies. There are social costs (including the impact on public health) and costs that fall on individual entities. To ease the "bumps in the road," emerges Green Trading or environmental financial trading.

THE FIRST WAVE

The tool, environmental financial trading, was invented in the United States. The first trades involved allowances for SO_2 (an air pollutant contributing to acid rain) under a national cap-and-trade program in 1995, followed by regional trading of allowances for NOx (nitrogen oxides that are precursors of smog) in 1999. There are now spot markets for SO_2 and NOx as well as over-the-counter (OTC) forward markets. The use of flexible market-based mechanisms has allowed the reduction of pollutants to fall below the emissions cap set by the U.S. Environmental Agency and at costs considerably lower than initially forecast.

This market-based solution has been adopted and extended by various European countries and by early corporate movers (such as BP, Shell, and DuPont) to trade carbon dioxide and other greenhouse gases in the European Union, the United States, Canada, and Australia. In moving ahead of the pack, these corporations were able to gain valuable trading experience in the new commodities and in the use of environmental financial hedging as a tool to mitigate exposure to environmental risks. Very importantly, these initial trades have helped to write the "rules of the game" and establish legal precedents, providing the most active countries with the opportunity to influence the development of international laws pertaining to environmental financial trading.

SIGNS OF THE SECOND WAVE

Environmental issues have evolved from the sole responsibility of the EHS (environmental, health, and safety) departments in companies as they become more intertwined with corporate financial concerns. The matter is now moving to the CFO level and framed as a financial corporate issue. Financial risk managers are more likely to be involved, given the need to disclose environmental financial risks on the company balance sheet under Sarbanes-Oxley, growing shareholders' concern, and increasing acceptance of socially responsible investing.

We also have begun to see frequent trades of options for carbon emissions reduction credits under the Clean Development Mechanism and in Europe, and these trades should increase due to the need to meet Kyoto commitments and E.U.-E.T.S. caps. At the same time, a significant number of financial companies have energy assets, exposing them to environmental risks. Therefore, it is a natural step for financial players to build GHG portfolios and take positions in the market for long-term gain. In fact, those that are long on carbon credits are beginning to warehouse such credits as their values appreciate over time.

DIFFERENT PATHS TO A CARBON-CONSTRAINED ECONOMY

With the Kyoto Protocol and E.U.-E.T.S. in place, the countries and companies involved must begin the real work of implementation to satisfy the agreed-upon goals. What will happen when the Kyoto commitment period ends in 2012 is unclear, posing uncertainties for long-term financial and investment planning. Leaders in the European Union have agreed to a 15 to 30 percent reduction in GHG emissions by the year 2020, subject to future analyses and negotiations.

While the United States, which is the world's largest economy and contributor of a quarter of the global GHG emissions, has opted to not join the

Kyoto Protocol, there has developed a groundswell of activity at the regional, state, and local levels. Indications of "bottom-up" efforts include:

- The power sector's voluntary efforts to reduce GHG emissions, such as Entergy's purchase of 1 million CO_2 emissions reduction credits from geological sequestration and Power Partners' Memorandum of Understanding with the U.S. Department of Energy, pledging to collectively reduce GHG emissions intensity (both in December 2004);
- The efforts of state attorneys general to establish carbon dioxide as an air pollutant, such as lawsuits against the U.S. Environmental Protection Agency and some of the largest utilities;
- The West Coast (California, Oregon, and Washington) Governors' Global Warming Initiative;
- California's proposed rule that would require the reporting of carbon dioxide emissions reduction associated with energy efficiency measures;
- The nearly 20 states that have or are developing Renewable Portfolio Standards;
- California's regulations mandating reduction of auto emissions, which may be adopted by several other states as well as Canada; and
- The National Commission on Energy Policy's recommendation for a mandatory, economy-wide tradable-permits system in 2010, aimed at reducing GHG emissions intensity.

In a 2004 survey[2] of electricity-generating companies, nearly 60 percent of the respondents believed that Congress will enact mandatory carbon dioxide emissions limits in the next 10 years and about half of the respondents believe that there will be such action within five years. Clearly, we are seeing an emerging recognition of the responsibility to set a cap on emissions. Such a cap will send economic signals for innovation and investment in low-emission projects and technologies. The United States clearly is moving toward mandated federal standards, perhaps much more quickly than many observers imagine.

Despite public perceptions that the United States is doing nothing on climate change, the truth is that much is being done. There is an active over-the-counter carbon trading market, 28 states are moving forward with greenhouse gas initiatives, there are carbon geologic sequestration efforts married to enhanced oil recovery, and a rising perception at the federal level that U.S. multinational corporations must comply with the

E.U.-E.T.S. and the Kyoto Protocol in other countries where they conduct business.

This book consists of essays by contributors, most of whom made presentations at the 2004 GreenTrading Summit™: Emissions, Renewables & Negawatts in New York City. The chapters, elaborating on the presentations, were written through the end of 2004. Collectively, they provide a window that shows the U.S. path toward a carbon-constrained economy and include examples from Australia, Italy, and Eastern Europe.

In Chapter 1, Peter C. Fusaro, Chairman of Global Change Associates, argues that the private sector will take the lead in developing emissions trading markets as environmental issues become corporate financial issues. He points out that, for these markets to grow, there need to be price indices and common, consistent metrics that are applicable across market segments of GHG emission reduction, renewable energy, and energy efficiency. Finally, he discusses the role of exchanges and the creation of the global CO_2 emissions portfolio.

In Chapter 2, Paul A. Hilton, Portfolio Manager at The Dreyfus Corporation, discusses climate change in the context of socially responsible investing. In evaluating a company, many analysts now consider the financial risks and rewards associated with climate change and they examine the corporate response to climate change at various levels of governance. Since such evaluation depends on good information, three efforts to increase transparency on climate change are described. Further, through dialogue and shareholder resolutions, social investment analysts have impressed on company management their concern for climate change as a corporate responsibility.

In Chapter 3, Sheila Slocum Hollis who chairs the Washington D.C. office of the law firm of Duane Morris LLP, describes the complex and contradictory trends that characterize the political landscape for environmental trading. She discusses the impact of the Kyoto Protocol on the intertwined energy/environmental systems and opines about the legislative stalemate over energy issues at the U.S. national level.

In Chapter 4, Susanne Haefeli and Einar Telnes of DNV Certification's international climate change services provide an assessment of the current global GHG markets and GHG credit prices. They discuss possible future scenarios in the political and technology arenas, including the role of renewable energy sources.

In Chapter 5, Richard L. Sandor and Claire M. Jahns of the Chicago Climate Exchange state that the overarching goal of CCX is to turn the theory of GHG emissions reduction and trading into reality—to prove that the concept can work in practice. They describe the features of CCX and some achievements during the first year of the Exchange's operation. Finally, they discuss the possibilities of extending the central exchange model to other environmental commodities and provide descriptions of the immediate real extensions: the Chicago Climate Futures Exchange and the European Climate Exchange.

In Chapter 6, Stefano Alaimo, who directs the Environmental Markets department at Gestore Mercato Elettrico (the Italian Electricity Market Operator), describes the mechanisms for green (renewable energy) certificates and white (energy efficiency) certificates as well as the operation of the Italian trading platform for these certificates. Black (GHG emissions reduction) certificates are related to the Assigned Amount of Units (AAUs) under Italy's National Plan in E.U.-E.T.S. and are traded at the European level.

In Chapter 7, Ed Holt, President of Ed Holt & Associates, discusses the expanding markets for RECs (renewable energy certificates) by drawing on three presentations at the 2004 GreenTrading Summit™. The first presentation on REC certification and verification was made by Karl Rábago, Energy Group Director at the Houston Advanced Research Center and Chairman of Green-e's Green Power Board. The second presentation on DuPont's voluntary renewable energy program was made by Ed Mongan, DuPont's Director for Energy and Environment. The third presentation on long-term support for RECs was made by Steven Weisman, Director of the Green Power Program at the Massachusetts Technology Collaborative. In providing an overall frame for these three pieces, Mr. Holt discusses several emerging issues in REC markets: a) the definition of the role of regulators; b) the tension between a desire for larger, more liquid REC markets and the desire for local benefits; c) renewable energy's limited access to the environmental markets; and d) the disaggregation of REC attributes.

In Chapter 8, Mark Little, Vice President of GE Energy's power generation segment, describes the historic roadblocks to the full harnessing of wind power and provides a picture of the evolution of the market for wind power. He discusses various factors driving the current wind power markets, the most compelling of which is the growing environmental pressure. Finally, he considers critical factors in the REC trading infrastructure for the successful monetization of RECs.

In Chapter 9, Dr. Mark C. Trexler, President of Trexler Climate + Energy

Services, provides a "Best Available Corporate Forecast" (BACF) approach for forecasting GHG prices. This approach recognizes that the GHG market is not a commodity market and that traditional commodity forecasting methodologies are ineffective, given the characteristics of the GHG market and the nature of its uncertainties. The BACF approach forecasts price ranges that can be meaningfully used by companies incorporating GHG planning into their strategic thinking.

In Chapter 10, Arthur Lee, Principal Advisor for Global Policy and Strategy at ChevronTexaco, describes the process used by his company to integrate GHG emissions management into its business planning. This chapter also discusses a portfolio of tools that could be used for this purpose in the planning and review of new projects.

In Chapter 11, John Palmisano and Deltcho Vitchev propose new financial and policy tools to promote sustained progress in energy efficiency and renewable power while simultaneously creating substantial environmental benefits, including GHG emissions reduction. They describe the concept of a proposed Energy Efficiency and Renewable Energy Fund for a select group of transitional economies, which they developed at the request of the United Nations Economic Commission.

In Chapter 12, Patrick R. Zimmerman provides a detailed description of C-Lock, a patent-pending Web-based carbon sequestration accounting and marketing tool. C-Lock, developed by Dr. Zimmerman and his colleagues at the South Dakota School of Mines and Technology, aggregates carbon emission reduction offsets from individual land parcels and prepares certified units for sale in the marketplace.

In Chapter 13, David G. Brand, Director of the New Forests Program at the Hancock Natural Resource Group, reviews the nature of institutional investment in forestry. Focusing on Australia's forestry sector, he considers the emergence of environmental markets and values as a factor in the evolution of investment.

In Chapter 14, Dick Kempka and Dawn Browne discuss the terrestrial carbon sequestration achieved through the land management and conservation activities of Ducks Unlimited (DU). Building on its relationship with landowners in the United States where 70 percent of the land is in private hands, DU aggregates credits associated with the carbon/GHG sequestered as a result of land restoration. As a result, DU can function as an offset provider, offering opportunities for energy companies to invest in its conservation projects as a way of offsetting power plant GHG emissions.

In Chapter 15, William G. Russell, CEO of SKN Worldwide, discusses selected technology trends that will enable complex sustainability-driven systems such as the environmental markets. He presents this discussion in the context of lessons learnt in the development of other technology-based markets.

Finally, in Chapter 16, we review the early-2005 landscape for the Green Trading Markets, including the beginning of shifts in thinking on Wall Street as well as the implementation of the Kyoto Protocol and the E.U.- E.T.S. The Triple Convergence that we foresaw is now taking shape as the second wave of Green Trading takes off.

---

[1] The six greenhouse gases are carbon dioxide, methane, nitrous oxide (N_2O), hydrofluorocarbons (HFCs), perfluourocarbons (PFCs), and sulfur hexafluoride (SF_6).

[2] PA Consulting, Washington, D.C.

Peter C. Fusaro, *Founder* and *Chairman* of Global Change Associates, an international energy and environmental consulting firm, can be reached at **T**: (212) 316-0223 and **E**: peterfusaro@global-change.com.

Marion Yuen is *President* of The MYA Group, a firm dedicated to creating and supporting partnerships that extend business competitiveness. She can be reached at **T**: (718) 230-5402 and **E**: myuen@mya-group.com.

The authors co-created the Annual GreenTrading Summit™: Emissions, Renewables & Negawatts (www.greentradingsummit.com).

Acknowledgments

We want to thank GE Energy and Tetra Tech, Inc., sponsors for the publication of this book, and Elsevier, the publisher. At the same time, we want to express our appreciation to the contributors who have shared their insights and experience in the various chapters as well as generously given their time and support.

Thank you to the many organizations and individuals who have enabled us to make the GreenTrading Summit™ an annual success. We want to acknowledge the individuals who have specifically helped us in the completion of this volume—Michele van der Westhuizen, Mary McCann, Alok Bhatnagar, John Borgersen, Karl R. Rábago, Michael Kujawa, David Brand, Hoi-Ying So, Heather McGeory, Kedin Kilgore, and Laura Tranter. Finally, we want to thank Stephanie Kloos of S. Kloos Communications, Suzanne Lander, Tita Young of Wordscribe.com, Sandi Schroeder of Schroeder Indexing Services, Ed Grant and Vicki Fischer for their help in the production of this book.

CHAPTER 1

Green Trading: Convergence of the Capital Markets and the Environment

By Peter C. Fusaro

The energy and agricultural industries—the world's leading emissions polluters—will be the leading suppliers of environmental solutions, because it is good business. Today, these industries are at a turning point on global warming as carbon intensity continues to grow while time to stabilize carbon dioxide and other greenhouse gas (GHG) emissions is limited. This issue goes far beyond the pitifully weak and flawed Kyoto Protocol on which many in the European Union (EU) and Asia have focused their efforts during the past decade. Time is *not* on our side.

The United States accounts for 25 percent of global GHG emissions. In order to meet growing electricity demand, this country (like the developing nations) is now moving to more coal-fired capacity with greater GHG emissions and other forms of pollution. At the same time, the U.S. government will never accept the Kyoto Protocol. So, rather than concentrating on the controversial issues of this treaty, it is now more important to focus on what can be done and how it can be accomplished. Trading and the markets offer a solution and a way forward.

The energy industry particularly has the financial strength, intellectual capital, and global presence to provide these solutions. BP and Shell have already taken the lead, but others are not far behind. ChevronTexaco has developed GHG software that it is sharing with the world (see chapter 10). Suncor, the Canadian tar sands producer, is now the biggest wind developer in Canada as it uses renewable energy as carbon offsets.

The carbon footprint of the majors can be found everywhere in oil and gas production, refining, and transportation around the world. As these companies continue to expand their involvement in the power industry, they will need to adopt solutions that include the use of more efficient, environmentally benign technology as well as basic changes in standard industry practices.

Another sector under the spotlight is electric utilities, which arguably has borne the brunt of responsibility in most GHG discussions. While some in this industry have been proactive in beginning remediation steps (noticeably, coal-burning AEP), many others are grappling with how to develop a GHG strategy.

MOVING BEYOND KYOTO

Since the private sector has a vested commercial interest in emissions reduction, it will take the lead on the development of emissions trading markets. Compliance responsibility, however, will rest with government. Prevalent is the strong belief that markets will form first and that government should not inhibit their growth.

European, Japanese, and U.S.-based companies are now moving ahead to develop both trading programs and pilot projects since a first-mover advantage exists, and waiting for regulatory approval may prove more costly in the future. Emissions rights may be traded through bilateral transactions or brokerage houses or by listing on exchanges.

The Kyoto Protocol envisions three international mechanisms that would enable Annex 1 countries to reach emissions-reduction targets beginning in 2008 through 2012. These mechanisms are emissions trading, joint implementation (JI), and the clean development mechanism (CDM), with all three modes currently being used.

Bilateral trade between countries is generally regarded as the most effective means to initially trade emissions. The emissions unit to be traded is 1 ton of carbon-dioxide equivalent (CO_2eq) for the 6 greenhouse gases.[1] Among these greenhouse gases, NOx and CH_4 (methane) emissions can be more difficult to quantify in many countries. The U.S. has already established an over-the-counter (OTC) market for NOx and has traded CO_2 emissions. In addition, efforts are underway to develop protocols for non-Kyoto CO_2 trades to be factored into a global trading market.

Since trading mechanisms will be part of any long-term approach to limiting GHG emissions, the emissions market is going forward on many fronts without Kyoto approval or U.S. participation in Kyoto. It is anticipated that actions taken today will most likely be grandfathered into a future revised treaty. Kyoto was meant to be flexible, allowing market-based solutions to trading GHG as a carbon-reduction strategy and as a means to facilitate the spread of energy-efficient technologies for industry. Further, since governments expect industry to make the largest GHG reductions, the obligation for progress falls heavily on the oil industry, electric and gas utilities, manufacturing, and automakers.

A RISING CORPORATE FINANCIAL ISSUE

Environmental issues are now becoming corporate financial issues. Greater financial disclosure of corporate environmental risks (including risks due to climate change) has raised the issue of the environment as a corporate fiduciary responsibility. Increasingly, the environmental and financial performances of companies are intertwined. This impacts automobile manufacturers, electricity utilities, hydrocarbon groups, banks, and insurance companies. Moreover, it is rising as a Sarbanes-Oxley issue as well, which means that environmental financial risks will now have to be disclosed on the company balance sheet.

Automakers are becoming concerned about carbon-dioxide emissions per vehicle, and utilities now pay more attention to cutting their GHG emissions as part of their overall air emissions reductions. Oil and gas companies are increasingly concerned about emissions as production, refining, transportation, and distribution liabilities. Banks' share valuations could fall if these financial institutions lack adequate carbon risk management strategies. Insurance and re-insurance companies are now at the forefront of confronting such financial hazards as catastrophic crop failures and epidemics of infectious disease due to climate change. These new financial liabilities for insurance and re-insurance companies could prompt them to drop coverage for certain companies, which will prompt change and market creation.

Environmentally-related corporate financial issues are now mobilizing shareholders to voice their concerns at annual meetings of the large oil companies, for instance. These shareholders cite studies such as those conducted by Innovest (the so-called "green Moody's") to show that companies perceived to be more environmentally aware are in fact more financially successful. Indeed, as corporations begin to analyse financial risks, they also realize that this global issue requires action. While the good deeds of BP, Shell, DuPont, Trans-Alta, and AEP are important, it is now time for the new wave of corporate engagement. Already, projects and trades have begun, much institutional money has flowed into project-based reductions, and green trading is now underway. For this second wave of corporate engagement, companies need the greenhouse-gas business case for taking action *now* and require the confidence that there will be no later penalty for such actions.

ENTER AGRICULTURE

The agricultural sector is beginning to realize the market potential and financial benefits of renewable energy—not just in the form of rents from siting large wind towers but from the more important self-generation of electricity with wind and biomass from agricultural wastes. The utilization of plant and animal farm waste can produce additional cash crops to be "harvested" and commercialized for their environmental attributes. The energy and agricultural sectors can join forces to develop new energy supplies while reducing externalities and creating new American industries that can be exported throughout the world.

Together, energy and agriculture are the world's largest businesses. Notably, they have also the most deeply liquid commodity markets. This liquidity provides excellent conditions for the financial engineering of environmental financial products that could capitalize on and grow cross-commodity arbitrage opportunities—not only for energy and agricultural commodities but also for GHG emissions reduction, renewable energy, and energy efficiency. The inflexion point for this sea change is during the next two years.

PROJECT FINANCE IMPLICATIONS

Another emerging financial trend that may hold the key to GHG emissions liquidity is the structured finance market, i.e., "Green Finance." A fuel-type shift to greener and cleaner fuels such as natural gas (in preference to coal or oil) is becoming embedded in the fabric of new power-station project financing. Since these plants have a useful life of 30 to 40 years, they will bring a stream of emissions credits that can be banked or used upfront, thus unlocking another avenue for market evolution. This type of thinking is just beginning to take hold at investment and commercial banks in New York, London, and Tokyo.

Moreover, it can be envisioned that an environmental checklist is emerging in the green or environmental finance arena—yet another way in which financial engineering can bring about market development and liquidity. There is no time to fight past demons. Forward-thinking and globally based energy participants should embrace the inevitability that international policy on GHG is being set by both media and public perceptions. In this context, the rational response by enlightened industry participants is to develop and support market-based solutions to global pollution.

In an imperfect world, this is the reality. In order to reduce or offset CO_2 emissions, emissions trading will act as a catalyst of change in the transition of world economies toward renewables and accelerated transfer of more efficient, greener technologies.

Ironically, the global market that now seems best positioned for trading is the renewable energy credit (REC) market. Renewable energy has undergone a quantum technology shift in terms of increased efficiency and lower costs, and there are only a few financial players focusing on the new factors that drive this market. Once again, government mandates (called *Renewable Portfolio Standards* in the U.S.) are driving market maturation. In the physical market, wind and solar power are posting global growth of 40 and 30 percent per annum respectively, with costs for wind power now competitive with gas and coal. In addition, tax subsidies for waste-to-energy and biomass power generation will move the equation further forward.

Looking at a small installed base of renewable power generation today misses the fact that the ramping up of this technology is global. These power stations are also getting bigger with wind turbines of 5.5 MW, and 300 to 400 MW wind farms on multiple sites are being developed. More important, they have created another fungible commodity market that can be traded across borders as the credits are measured in megawatt hours. Such green power initiatives will create a highly fungible market for RECs.

THE EMERGING FINANCIAL MARKET

Energy trading began in 1978 with the first oil futures contract on the New York Mercantile Exchange (NYMEX). During the 1980s and 1990s, the International Petroleum Exchange (IPE) and NYMEX successfully launched futures contracts for oil and gas. These successful futures exchanges survived the Enron et al. energy-trading debacles of recent years and demonstrated their capable financial performance. Today, oil companies and financial houses provide the necessary trading liquidity through market-making on both the established government-regulated futures exchanges and off-exchange energy derivatives markets, which can clear on the futures exchanges. These companies know how to manage their financial energy risks and have the risk-management skills that will be deployed increasingly in the emerging global environmental markets. Financial risk will be managed on established energy futures exchanges because trading debacles have taught the energy markets that financial performance is fundamentally important. While OTC brokers (such as Natsource, Evolution Markets, and CO_2e) broker bilateral trades, market-making is what is lacking from the environmental financial markets. However, in order to make a market, principals are needed.

The principals for environmental financial market-making will be the investment banks, multinational oil and gas companies, and agribusiness. They have the global presence, balance sheet, and the exposures to take action and to put their financial wherewithal behind this market as they have done for oil and gas trading. They also have the financial balance sheet to perform.

Environmental financial products for sulfur dioxide (SO_2) and nitrous oxides (NOx) have been successful in controlling U.S. pollution since 1995. A $6 billion environmental market today may seem pale in comparison with a $2 trillion energy derivatives market, but the growth trajectory suggests that today's green trading markets should be compared with 1978's oil markets. However, this time around, maturation will be global and simultaneous as carbon-trading regimes take root in the EU, Asia, Australia, and North America. While thus far, trades for carbon dioxide have numbered only in the hundreds—with a notional value of about $500 million—estimates suggest that a $3 trillion commodity market may emerge over the next 20 years. The dollar value of this market is enticing, but the reality is that the global energy industry will be one of the primary suppliers of liquidity to this market, followed by the agricultural industry, since both industries are already active in commodity trading.

Green trading encompasses the convergence of the capital markets and the environmental markets; it includes not only trading in GHG emissions reduction but also renewable energy and the financial value of energy efficiency. Further, there are natural cross-commodity arbitrage opportunities since oil, gas, coal, and power, like weather derivatives, have environmental dimensions. Today, cross-border trades of carbon dioxide have been conducted between the U.S. and Canada, Canada and Germany, Germany and Australia, and Australia and Japan. Developing countries will be fully engaged in this financial market as sellers of GHG credits and allowances, using its mechanisms to provide liquidity for needed technology transfer.

Green trading provides a market-driven solution to reduce pollution, but government sanctions are needed to put the rules in place. The U.S. SO_2

program is a "cap-and-trade" plan with a 35-year life that requires the retirement of pollution credits from 1995 though 2030. A GHG regime will require a 100-year life and should be put into place now, not in 15 years. Governments must also deal with the cross-border components of trading, and rules need to be harmonized. As in the overall environmental financial market, liquidity providers in the green-trading markets will include energy companies, banks, agricultural producers, insurance and reinsurance industry, and investment banks.

U.S. EMISSIONS TRADING EXPERIENCE

Although many countries continue to propose various types of emissions-trading initiatives, the reality is that only the U.S. has the track record of a successful emissions-trading market which has worked well over the past 10 years. As initially proposed by the Environmental Defense Fund (a U.S. environmental organization now called Environmental Defense) to the first Bush administration for the trading of sulfur dioxide (SO_2) credits, the emissions-trading market has been successful even beyond what its architects envisioned.

Basically, during March of each year, the U.S. Environmental Protection Agency (EPA) runs an emissions auction supervised by the Chicago Board of Trade. Under Phase I (which began on January 1, 1995), the 110 highest emitting utility plants were mandated to reduce their annual sulfur dioxide emissions by 3.5 million tons. This process began in 1995 for sulfur dioxide and was extended to nitrous oxides (NOx) in 1999. The OTC forward markets trade these vintaged credits through the year 2030. (Vintages are credits available for sale each year until they expire.) Several OTC energy brokers (including Evolution Markets, Natsource, Prebon, and Cantor Fitzgerald) are involved in brokering these credits, and over one million trades per year occur. Thus, the market is liquid and has created emissions credits that are a fungible financial product. It has also saved $1 billion per year over command-and-control strategies of the past. Under Phase II (which began on January 1, 2000), a more stringent standard called for an additional annual reduction of 5 million tons of sulfur dioxide, and the program was expanded to another 700 utility plants throughout the U.S. Today, that financial market is indicating SO_2 prices of over $500 per ton, which creates financial incentives to reduce pollution.

Under the SO_2 program, utilities are given flexibility on how they meet the mandated targets, and can switch to fuels with lower sulfur content, install pollution control equipment, or buy allowances in order to comply with the law. The utilities are given one allowance for each metric ton of sulfur dioxide emitted. These emissions allowances are fully marketable once they are allocated through an EPA auction, and can be bought, sold, and banked. In order to sell allowances, utilities must reduce their emissions below their emissions limit. All transfers are recorded in the allowance trading system and posted on the Internet. Serial numbers allow the tracking of each allowance's trading history, and an inventory for all accounts is available.

The allowances are allocated in phases. The later phases tighten the limits on previously impacted sources of pollution and are also imposed on smaller, cleaner units. Compliance is assured through continuous emissions monitoring at plants and regular reports to the EPA. Fines are assessed if companies don't comply with the law. Learning from this successful experience, mandatory standards will also be needed for CO_2 reductions as the value of voluntary compliance has been currently valued at $2 dollars or less per metric ton.

THE NEED FOR PRICE INDICES

Markets in environmental financial derivatives are positioned for rapid growth due to political initiatives and business opportunities, but these markets will reach their full potential only if based on reliable indices widely accepted by the trading community. To focus solely on GHG emissions misses the opportunity to capture the benefits of other energy/environmental market-based solutions to global pollution such as renewable energy credits or energy efficiency (negawatt) trading. Therefore, in order to maximize the business opportunity for an established exchange, several environmental financial products for various geographic markets must be traded using regional environmental indices as the underlying benchmarks. The composite of these financial indices will contribute to a global index as well. The need is to establish exchange-traded derivatives products for sulfur dioxide (SO_2), nitrous oxides (NOx), carbon dioxide (CO_2), renewable energy credits (RECs), negawatts (energy efficiency), mercury, and other environmental verticals, with the first step being the creation of several tradable indices in North America, Europe, and Asia.

Since government mandates are the primary market driver for environmental financial products, the scope of activity has been limited to a small number of players. Due to the lack of mandatory compliance for carbon reductions, there are still more sellers than buyers. Nonetheless, the growth of emissions trading and profit opportunities are attracting a new generation of traders in the market. Commodity traders from the world's largest banks and financial institutions are responding to these opportunities by opening trading operations on both sides of the Atlantic.

One inhibiting factor to market development is the lack of reliable and liquid financial indices, which has muted efforts to create a liquid market. The current trading environment is handicapped by the operational complexity of having adequate allowance inventory on hand to complete a trade, which limits access only to those with ample allowances or those that can borrow allowances. Furthermore, the process of transferring allowances from one party to another can take weeks, limiting traders' ability to enter or exit the market with ease. An index would remove this impediment by allowing more trade structures and by turning the environmental market into a cash-settled operation with the added benefit of improving cap-and-trade policy. Consequently, this would attract more players into the market. With the September 2004 launch of a clean technology index at the American Stock Exchange, it is not too farfetched to expect the appearance of a variety of environmental indices in the not-too-distant future. Because of the potential for improving regulatory policy, we would expect close cooperation between government regulatory agencies and any exchange seeking to use the indices as underlying benchmarks for trading financial products.

NEED FOR NEW AND CONSISTENT METRICS

Many countries have renewable energy, energy efficiency, and GHG programs. Since most programs today are and have been independently developed, there needs to be some coordination to provide consistency. Consistent methodologies for measuring emissions—including GHG, renewable energy, and energy efficiency efforts—would facilitate project investment. Consistency would facilitate development of project templates, thereby reducing costs and allowing rapid dissemination of the lessons from early projects. National and international markets for GHG credit trading would offer the liquidity necessary to return value to projects and thereby, encourage financing. To function efficiently, such markets require assurance of integrity—clear definitions, avoidance of double counting, verification, and liquidity. At this point in market development, it is critical to foster some consensus around the development of common metrics for the private sector and policy makers to assess opportunities at the regional, national, and international levels. GHG registries managed by a third-party, non-governmental entity could serve as a model at a state or federal level (such as in California and most EU countries, respectively).

Today's one-off market is composed of many companies not acting on what will ultimately help them financially. A few innovators are proactive. The reality is that environmental damage is emerging as a financial liability for multinational corporations globally. These liabilities are the market drivers for change. As the dynamic models have yet to be built, the quantification of these risks will keep analysts and mathematicians busy for many years.

Software products for both quantifying forward prices for CO_2 and RECs will be very valuable to a host of users. Demand for such software products has been stimulated by the January 2005 launch of the EU Emissions Trading Scheme (ETS) as well as increased price volatility in the U.S. SO_2 markets due to increased oil price volatility and higher prices, which has lead to a knock-on effect on SO_2 markets.

2005: A BREAKTHROUGH YEAR

The year 2005 could be the breakthrough year for this emerging financial market. Besides the vaunted EU ETS, there is significant movement at the state level in the U.S. More than nine states are collaborating in the Regional Greenhouse Gas Initiative (www.rggi.org) to form a cap-and-trade market in the Northeast, which will be in harmony with Canadian provincial governments' requirements in eastern Canada. This initiative also has an agreement to work with the California Climate Action Registry (www.climateregistry.org) to have conforming standards. In 2005, these developments would set in place rules to begin cross-state GHG trading in the U.S. Moreover, after the 2004 presidential election, the federal government probably will seek standardization to ensure harmony between the U.S. energy industry and others as well as overseas administrations because U.S. corporations need this certainty for investment planning. Japan is not far behind as it undertakes mock trading of carbon with over 40 industrials.

There will be two stages in the development of the international carbon market. Now, in stage one, carbon credits are being created. Trading covers many years because, thus far, there has not been an allocation of sufficient units to have a spot market and because the units are project-based reductions. Capital is required, and forward commitment for carbon-dioxide reductions cannot be banked at the present time. If the World Bank were buying a 10-year stream of reductions, a bank loan would usually be available to implement the project.

Consequently, trades are still done in large-volume structured deals. Nevertheless, early speculative trading and some risk hedging have begun. Alongside this emergence, there is a transformation in how climate change liabilities are handled within some energy companies and energy end-user groups. As major corporations begin to treat the GHG issue as a financial matter, responsibility is passing from environmental professionals to risk managers. In this, the early stages of the market, carbon finance is playing a bigger role and, over the next years, a liquid spot market will develop.

Green trading markets are now at a turning point. The existing market is characterized by opaque prices, little trading, few participants, poor liquidity, tremendous inefficiency, and wide arbitrage opportunities—factors that brokers now love; these attributes are familiar because they occur in every new market during its market maturation process. Having seen the emergence and maturation of oil, gas, power, weather, and coal as fungible commodity trading markets, the environment is now well positioned to be the next financial commodity trading market.

Uniquely, the carbon market will develop simultaneously throughout the world—something that has never occurred in other markets. The second stage of carbon market development will be toward a mature and liquid market and, over the next 10 years, there will be linked markets followed by indexed markets. We shall see spot trading, high volumes, advanced brokerage similar to the power and gas markets, and a growth in carbon finance.

Moreover, another unique aspect of this market is that it will be a government-mandated market despite advocates of voluntary trading in the U.S. Arguably, the U.S. created the carbon template: The trading regime of the sulfur dioxide (SO_2) allowance market (which began in 1995), as described earlier in this chapter, has vintage credits to the year 2030 while a true carbon regime will have a span of 50 to 100 years. This is envisioned after 2012 for the Kyoto Protocol, and work at the governmental level has begun to create the longer-term market.

This new marketplace would motivate firms with surplus emissions rights to trade or supply those rights to the market. Despite the risk of uncertainty on future rules, there are advantages in early action. The argument today is that to do it early will probably be less costly than in the future. Accumulating GHG emissions allowances now is a form of insurance for industry participants. Moreover, emissions trading delivers significant environmental benefits from reduced compliance costs as well as promotes environmental technologies. As rules become more clearly defined, industry-driven schemes will probably play key roles and be grandfathered into future regimes. Thus, industry can create its own domestic and international portfolio of emissions allowances or credits.

Emissions trading schemes have various characteristics similar to the dual process of electric power industry liberalization in many countries. Since the power industry contributes substantially to GHG emissions, the intersection of emissions trading and electric power deregulation will provide impetus to move the process forward.

THE FUTURE ROLE OF EXCHANGES

Almost all environmental financial contracts, such as those in SO_2 or CO_2, are traded on the OTC markets. Therefore, there is an opportunity for exchanges such as the IPE, NYMEX, and the Chicago Board of Trade to offer OTC clearing, which would effectively make these quasi-futures contracts under government oversight and help make them more acceptable to risk managers. The IPE recognized this opportunity last April and has linked its platform to the Chicago Climate Exchange in order to trade emissions in the EU.

In Japan, both the Tokyo Commodities Exchange and Tokyo Stock Exchange are considering launching carbon derivatives contracts. Presently, the ground rules in Japan are in a state of flux between a cap-and-trade market and a baseline market. A movement is also emerging to create the next trading regime beyond 2012 and for the Kyoto Protocol to include developing giants such as China, India, and Indonesia.

In launching a voluntary carbon exchange in September of 2003, the Chicago Climate Exchange or CCX (www.chicagoclimateexchange.com) is following another route to GHG market maturation. This voluntary carbon exchange is mostly U.S.- and Canada-centric, with a current roster of more than 60 members. As the first exchange to be launched at a time of changing U.S. attitudes on global warming, it serves as a precursor to other North American exchanges that are likely to enter this emerging market space. As the next step, CCX has partnered with the IPE to launch the

European Climate Exchange in time for the EU ETS in January 2005.

There is competition to create global environmental exchanges. In actuality, the exchanges need not be mutually exclusive as today's Internet technology facilitates borderless trading. In effect, we can have world GHG trade through the Internet. Because exchanges can be established quickly on the Internet, many believe that Internet-based emissions trading would be a desirable development. Such trading would have low costs of operation and allow immediate disclosure for market players. The concept behind the allowances was to foster the implementation of demand-side efficiencies or use of renewable energy. These concepts are tailored to the developing CO_2 market development and the use of the Internet as the means to implement change.

CREATING THE GLOBAL CO_2 EMISSIONS PORTFOLIO

The goal is a gradual reduction in emissions driven by measurable targets using market-based incentives that can include outright purchase of emissions reductions. The aim is to encourage better technologies, fuel choices, and results, and accelerated technology transfer. Already, multinational companies in North America, Europe, and Asia are developing emissions reducing initiatives that can be transferred to their affiliates in developing countries. In coming years, global environmental corporate portfolios will be managed with stringent profit and loss targets for company business units and will need fungible markets with price certainty in order to benchmark their financial performance.

Any market needs trading liquidity in order to ensure fungibility. So far, the CO_2 emissions trading market has completed about 200 trades, and its development is dependent on the resolution of many factors, not the least being caps. Since this market is in its infancy, trading caps can either be adopted by government or left open-ended for the markets to decide.

FUTURE MARKET DEVELOPMENTS

Green trading promises to be a $3 trillion commodity market involving major energy company participation. It will have cross-commodity arbitrage opportunities with oil, gas, power, and coal futures contracts as well as OTC contracts. It will create new project development in the renewable-energy and energy-efficiency sectors that will trade their environmental attributes. The global dimension of all these implications cannot be understated: Green trading will be the first truly global commodity market since the development of the oil market, and the coming years will see an acceleration of this market's maturation.

Peter C. Fusaro, *founder* and *chairman* of Global Change Associates, an international energy and environmental consulting firm, can be reached at T: (212) 316-0223 and E: peterfusaro@global-change.com.

---

[1] The six greenhouse gases addressed by the Kyoto Protocol are carbon dioxide (CO_2), methane (CH_4), nitrous oxide (N_20), hydrofluorocarbons (HFCs), perfluorocarbons (PFCs), and sulfur hexafluoride (SF_6).

CHAPTER 2

Carbon/Greenhouse Gas Transparency and Socially Responsible Investing

By Paul A. Hilton

With every passing day, institutional investors in the United States are growing more interested in the issue of global climate change. This increased attention is fueled by greater understanding of climate change as a reality, the financial risks associated with climate change, and the differences emerging in how companies are managing climate risk.

This issue has particularly hit a nerve with the subgroup that actively works on socially responsible investment (SRI) products at a number of U.S. investment firms. These analysts have an explicit mandate to consider social and environmental information as they make stock selections. For these investors, climate change has quickly emerged as the primary example of an environmental issue that has clear links to material financial risk.

Analysts now have the opportunity to examine the various corporate responses to climate risk in order to identify leading companies. Even though companies in different sectors have dissimilar levels of exposure to climate risk, a proactive response to this issue serves as a good proxy for overall quality of management for energy and retail companies alike. Companies that develop a strong management response to climate change, track Greenhouse Gas (GHG) emissions, set goals for reduction, and actively communicate their progress to investors and the general public, to name just a few examples, have an opportunity to distinguish themselves from their peers and attract investment.

EVOLVING U.S. SOCIALLY RESPONSIBLE INVESTMENT

According to a 2003 study by the Social Investment Forum, socially responsible investing in the U.S., broadly defined, now represents some

$2.16 trillion in assets.[1] SRI in this country has always been closely linked to the practice of avoidance screening, elimination from a portfolio companies that conflict with the investor. More recently, however, U.S. SRI is evolving in two important directions:

1) **Integrated Company Analysis.** Analysts are beginning to integrate social and environmental analysis into financial analysis, using such qualitative information as an extra lens to uncover potential financial risk and opportunity.
2) **Company Engagement.** Investors are leveraging their position as active shareholders to promote change through direct dialogue with companies and/or filing shareholder resolutions.

The issue of climate change represents an ideal way to view this evolution on both fronts.

CLIMATE CHANGE AND INTEGRATED COMPANY ANALYSIS

General Links between Social, Environmental, and Financial Performance

Research analysts who work on SRI products often observe links between social or environmental factors and financial results. Large environmental fines or corrective regulatory actions have led to quarterly charges and operating disruptions that reduce earnings targets. Major product safety issues have resulted in decreased product sales as well as high settlement costs. Strong workplace programs result in lowered absenteeism and turnover, in addition to higher worker productivity. Even more important, each of these examples directly translates to impacts on management credibility and overall corporate reputation, characteristics that mainstream Wall Street analysts frequently tout. Increasingly, social analysts are actively looking to find these links to help inform the stock selection process.

Links with Climate Change

Enhanced sensitivity to the interplay between social, environmental, and financial factors has allowed SRI analysts to earlier understand the potential risk implications of climate change. Risk to companies may take the form of direct impact from extreme weather events and climate shift: damage to physical assets, increased insurance claims, or loss of core business (think: skiing and wineries). The risk may also be indirect, such as through costs of possible legislation of carbon and other emission reductions. Most truly global companies will feel pressure in some way to conform to the terms of the Kyoto Protocol, or they will be variously impacted by the E.U. Emissions Trading Scheme, other national initiatives, and pending state or regional level actions in the U.S. (whether legislation or litigation). Electric utilities, in particular, could face enormous, but highly differing costs in a carbon-constrained world, primarily based on their relative reliance on coal as a primary fuel source.

Companies that make a point of qualifying reductions now for trading credit could reap financial benefits, as could companies that focus on cost savings tied to strong energy management. Retail companies, for example, have little pricing power and are forced to focus on better energy management as the price of electricity soars in some markets (electricity use by business is a primary contributor to U.S. GHG emissions) [see Table 1]. Given these likely risks, and possible rewards, many SRI analysts believe that consideration of this issue in company evaluation is part of their obligation as a good fiduciary. As Innovest writes, "To fulfill their fiduciary duties, investors and directors now must understand which industry sectors and companies are exposed to the greatest [climate change] risks, what measures are being taken if any to reduce them, and how effective they are likely to be."[2]

Table 1

Analysts are beginning to digest the financial risks and rewards associated with climate change

Extreme weather events/ climate shift	➡	Increased insurance costs (coverage and claims), damage to physical assets, loss of core business
Probable regulation	➡	Emissions reduction costs
Possible litigation	➡	High legal and settlement costs
Emissions trading	➡	Profit opportunity from early reduction and trades
Energy use management	➡	Lower operating costs

The Corporate Governance Connection

The raft of corporate scandals over the last few years has also turned SRI analyst attention to corporate governance reform. These analysts now see that companies must have leading corporate governance practices in order to promote better handling of important social and environmental issues. Certain structural changes are needed to promote improved performance in these areas, such as board independence, board diversity, annual election of directors, board committee oversight of governance and policy issues, and formal mechanisms for corporate social responsibility (CSR) professionals at the company to report to the board on a regular basis. This new focus has allowed the SRI field to forge alliances with other stakeholders who have been pursuing governance reform for some time. *Barron's*, a leading financial magazine read by many on the Street, acknowledged this trend in July 2003:

> For years, "socially responsible" investors were derided by many Wall Streeters as muddle-headed leftists or hopeless do-gooders. But the tidal wave of disclosures about wrongdoing in Corporate America's executive suites and boardrooms has won this group important allies, including pension funds, unions, and individual investors concerned about corporate governance.[3]

The 2003 CERES report, *Corporate Governance and Climate Change: Making the Connection,* makes the case that proactive steps taken by companies on climate change should be seen within a governance context.[4] In this report, author Douglas Cogan of the Investor Responsibility Research Center (IRRC) provides a checklist of 14 actions a company can take to address climate change, beginning at the top with board level changes. These actions offer a useful starting point for analysts seeking to differentiate company performance on this issue [see Table 2]. The report also profiles twenty high-GHG-emitting automotive, energy, industrial, and electric utility companies, identifying which of these actions each company has taken. Cogan is currently working on a more detailed breakdown of these fourteen steps that would allow for companies to be evaluated on a 100-point scale.

In addition to these fourteen points, analysts might consider a few additional key indicators to judge if a company is responding fully to potential climate risk. First, analysts should consider a company's com-

Table 2

Climate Change Governance Checklist

Board level:

1. Assign a committee of directors with direct oversight responsibility for environmental affairs.
2. Conduct a formal board-level review of climate change and monitor company response strategies.

Management level:

3. Place the chief environmental officer in a position to report directly to the chief executive officer of the CEO's executive committee.
4. Make the attainment of greenhouse gas targets an explicit factor in employee compensation.
5. Have the CEO issue a clear and proactive statement about the company's climate change response and greenhouse gas control strategy.

Reporting:

6. Include statement on material risks and opportunities posed by climate change in the company's securities filings.
7. Issue a sustainability report based on the Global Reporting Initiative or comparable "triple bottom line" format, which includes a discussion of climate change and a listing of the company's greenhouse gas emissions and trends.

Emissions data:

8. Calculate and register greenhouse gas emissions savings or offsets from company projects.
9. Conduct a system-wide inventory of the company's emissions and report the results directly to shareholders.
10. Establish an emissions baseline (dating back at least ten years) by which to gauge the company's emissions trends.
11. Make projections of future emissions and set firm, companywide targets to manage and control them.
12. Hire a third-party auditor to certify there are no material misstatements of the company's emissions data.

Other actions:

13. Participate in an external voluntary greenhouse gas emissions trading program.
14. Purchase and/or develop renewable energy sources.

Source: CERES Report, *Corporate Governance and Climate Change: Making the Connection,* prepared by Doug Cogan at IRRC, June 2003.

mitment to energy management: Does a company disclose energy management programs, set energy-use reduction goals, and participate in U.S. EPA's Energy Star program? Second, analysts should assess if companies in certain sectors have included an analysis of product contributions to GHG emissions (an important issue for automotive manufacturers and oil companies, for example). Third, analysts should mark whether companies participate in voluntary programs to address climate change, such as EPA's Climate Leaders program for tracking and reporting GHG emissions. Analysts looking to identify the relative exposure of companies in specific sectors can gain insight from new reports on climate change published over the last few years by firms such as Innovest Strategic Value Advisors,[5] West LB,[6] and Sustainable Asset Management.[7] This research not only outlines general climate change risks and opportunities, but also differentiates company exposure in specific sectors, such as automotive and electric utilities.

The Need for Disclosure

The greatest challenge for investors who grapple with evaluating climate risk is access to good information about company exposure and response. Three notable efforts to increase transparency around this issue are: The Investor Network on Climate Risk, the Carbon Disclosure Project, and the Global Reporting Initiative.

In November 2003, CERES (Coalition for Environmentally Responsible Economies), a Boston-based coalition of investor, environmental, labor and public-interest groups focusing on environmental responsibility, led an Institutional Investor Summit on Climate Risk at the United Nations in New York City. At this meeting, thirteen representatives of state pension funds with over $1 trillion in assets came together with investment professionals and world leaders, such as Al Gore and Kofi Annan, to discuss the risk of climate change. From this initiative, a group called the Investor Network on Climate Risk (INCR) formed. In April 2004, pension fund leaders, including the pension fund heads of California, Connecticut, New York State, and New York City, called on the U.S. Securities and Exchange Commission (SEC) to require disclosure of the financial risks of climate change in securities filings.[8]

Another investor effort to promote transparency on climate change, The Carbon Disclosure Project, represents 95 institutional investors with over $10 trillion in assets under management. This initiative has twice asked companies in the FT500 Global Index, most recently in November 2003, to complete a questionnaire about their response to climate change, including potential risks and opportunities. Innovest authored a report on the findings of this inquiry, *CDP2,* in May 2004 that found the number and quality of responses were up since the prior report, with questionnaires completed by 59 percent of companies, up from 47 percent.[9] Companies with "best in class" responses are included in an index of the top 50 companies across 12 sectors, from integrated oil and gas to diversified financials. Similarly, companies that did not respond fully to the information request are clearly identified.

The Global Reporting Initiative (GRI) seeks to encourage companies around the world to use a standard framework for reporting on sustainability issues. The GRI provides a set of Sustainability Reporting Guidelines that includes core reporting principles and key indicators. While uptake in the U.S. has been slow to date, a growing number of U.S. companies have taken the time to at least reference the GRI as they produce their corporate responsibility reports. The list includes 55 U.S. companies, including Pepsi, IBM, GM, Ford, Chevron Texaco, Intel, and Citigroup.[10] Only Ford and GM are "in accordance" reporters, committing to a higher level of reporting, which includes all indicators unless a reason is given for a specific omission. As key indicators of overall sustainability performance, GRI asks for disclosure of direct and indirect greenhouse gas emissions and energy-use figures.

As disclosure improves, so will analysts' ability to make good distinctions between corporate performance on this issue and relative risk exposure.

CLIMATE CHANGE AND COMPANY ENGAGEMENT

Dialogue

Increasingly, social investment analysts are using their access to companies to raise a host of corporate responsibility issues. While groups like the Interfaith Center for Corporate Responsibility have led company dialogues on CSR issues for years, the research analysts are using company meetings to inform their securities selection process. For example, in New York City, SRI analysts from The Dreyfus Corporation, Neuberger Berman, Citigroup Asset Management, Rockefeller and Co., Domini Social Investments LLC, and Christian Brothers Investment Services, Inc., come together on a regular basis to have meetings with companies on a variety of CSR issues. In the last year they have met with high-level representatives from Exxon Mobil, Staples, BP, Intel, and Microsoft, to name a few. Colleagues in other locations participate in the New York City meetings in person or by phone, or host their own discussions in cities such as Boston or Washington, D.C. At any one meeting, $10 billion in SRI assets might be represented, often with significant holdings of the company presenting. Companies have an opportunity to outline their CSR programs, receive feedback on their reporting, and get alerted of new "hot button" issues coming to the forefront. Typically climate change is at the top of the list of issues discussed by this group.

Shareholder Resolutions

In addition to dialogue, or often as a means to initiate dialogue, some social investors file shareholder resolutions with companies in an effort to push for changes in social and environmental policies or reporting. These resolutions come to a vote by all shareholders at company annual meetings. While typically shareholder resolutions on social or environmental issues do not receive majority votes, even a significant smaller vote can send a strong message to management and induce a change. At times, companies agree to provide concessions before the resolution comes to a vote in order to have the resolution withdrawn. For the 2004 proxy season, 28 shareholder resolutions on climate change were filed at 22 companies. While in 2003, five companies received more than a 20 percent vote of support (AEP, TXU, Southern, GE, and Exxon Mobil), already this year three companies have reached this level of support, including 27 percent at Marathon Oil, 28 percent at Apache Energy, and 30 percent at Anadarko Petroleum.[11]

Utility companies AEP and Cinergy have negotiated resolution withdrawals by agreeing to issue reports outlining the risks associated with the issue. Reliant also had its resolution withdrawn after it agreed to increase disclosure of environmental issues on its Web site and in financial documents. TXU and Southern Company agreed to issue a report on the companies' potential response to different climate-change scenarios. One company, Devon Energy, has negotiated a withdrawal of a resolution by agreeing to conduct an inventory of GHG emissions and develop an annual report.

CONCLUSION

Climate change is already an important factor for funds with a mandate to consider social and environmental issues. While mainstream financial analysts are beginning to gain awareness, few are factoring climate risk into their financial models. Those analysts who look at this issue first gain a competitive advantage by spotting early companies with lower risk exposures and higher potential returns. Proactive companies may also ultimately benefit from improved financial valuations relative to peers. The challenge now is for these companies to follow with full disclosure, so that analysts who are tracking climate risk can distinguish the leaders on this important issue.

[1] *2003 Report on Socially Responsible Investment Trends in the United States.* Social Investment Forum. www.socialinvest.org.

[2] *Value at Risk: Climate Change and the Future of Governance.* Innovest Strategic Value Advisors, Inc. April 2002.

[3] "Good Vibes: Socially Responsible Investing Is Gaining Fans . . . and Clout." Robin Goldwyn Blumenthal. *Barron's.* 7 July 2003.

[4] *Corporate Governance and Climate Change: Making the Connection.* CERES report prepared by Doug Cogan at IRRC, June 2003.

[5] *Value at Risk: Climate Change and the Future of Governance.* CERES and Innovest Strategic Value Advisors, Inc. April 2002. And, with WWF, *Power Switch: Impacts of Climate Change Policy on the Global Power Sector.* Dr. Martin Whittaker, Mark Kenber, and Rebecca Eaton. 17 November 2003.

[6] *Carbonomics: Value at Risk through Climate Change.* Dr. Hendrik Garz and Claudia Volk. West LB. July 2003.

[7] *Changing Drivers: The Impact of Climate Change on Competitiveness and Value Creation in the Automotive Industry.* Duncan Austin, Niki Rosinski, Amanda Sauer, and Colin Le Duc. SAM (Sustainable Asset Management) and World Resources Institute. 2003.

[8] "Thirteen Pension Leaders Call on SEC Chairman to Require Global Warming Risks in Corporate Disclosure." CERES press release, 15 April 2004. www.ceres.org

[9] *Carbon Disclosure Project: Climate Change and Shareholder Value in 2004.* Innovest Strategic Value Advisors, Inc. May 2004.

[10] www.globalreporting.org.

[11] "Global Warming Resolutions at U.S. Oil Companies Bring Policy Commitments from Leaders, and Record High Votes at Laggards." CERES press release, 28 April 2004.

Paul A. Hilton is a *portfolio manager* for socially responsible investing at The Dreyfus Corporation. He can be contacted at **T**: (212) 922-6292 and **E**: hilton.pa@dreyfus.com.

CHAPTER 3

The U.S. Political Landscape and Its Impact on Environmental Trading

By Sheila Slocum Hollis

The world is faced with sharply conflicting philosophies and political forces affecting the development, transportation, utilization, and conservation of energy resources. This presentation encapsulates the numerous and often contradictory trends that form the matrix underpinning contemporary energy law which ultimately affects any initiatives.

Following my executive service in government 22 years ago, I co-authored a book entitled *Energy Decision Making*[1]. The book was written during a period of high energy prices, blackouts, shortages, embargoes and turmoil in the Middle East, the full invigoration of a powerful OPEC, widespread market manipulation, and a time of geopolitical struggle for energy resources. Also, profound concern about the world's ecology was crystallizing. *Energy Decision Making* contained many comments that hold true today. I wish to quote a few lines and ask "what have we learned?" Oil prices are at an all-time high and international political instability plagues the energy markets. The environmental concerns have only expanded. History, in other words, is repeating itself.

> Energy, and the issues of its availability and price, have been prominent in the public consciousness...Prior to the... oil embargo in October 1973, the only publicly discerned or recognized central-government policy regarding energy was to assume the continued supply of cheap, abundant resources to feed the needs of the energy-hungry country. About 1970, policymakers awakened to the possibility that domestic supplies of cheap, readily obtainable energy were disappearing. To supplement declining domestic supplies

> and satisfy increased national demand, the nation became increasingly dependent on foreign-energy sources. That dependence, although reduced, continues today, and our government's energy decisions during the past decade directly reflect perceptions of the magnitude of that dependence.[2]

These were my words then, but where are we now? OECD countries together with vastly expanded world populations and economies have placed even more pressure on a finite supply base.

> Since 1970, the publicly perceived scarcity of domestic supplies and deep concerns about the security of the traditional foreign-energy-supply sources, together with severe doubts about the stability of the international energy supply-and-demand pool, have provided the fertile base for the growth of the role of the federal government in the control of energy policy. Although the Reagan administration has attempted to shift attention away from the so-called energy-scarcity mentality and from fear of reliance on foreign-energy sources, the energy issues that precipitated strong regulatory reactions in the Nixon, Ford, and Carter administrations have not suddenly evaporated.[3]

I believed that these issues were artificially dormant and added that if any major supply interruption from foreign sources were to occur, or if demand for energy increased significantly and more rapidly than predicted today, the federal government would be forced to respond to chaos. This is a reality that the administrations of Bush I, Clinton, and now Bush II have accepted while knowing that chaos could revolve around the equitable distribution of diminished oil supply and the economic adjustments consequent to vast price increases in a supply-constrained environment. Not only have we been grappling with that potential via attempts at workable energy policies, but also there have been two wars and multiple skirmishes with clear ties to concerns vis-à-vis the availability of oil supply to the world and OECD countries in particular.

I wrote that the belief that the marketplace can resolve every problem including the critical supply shortage is misplaced. Since the embargoes of the 1970s, the U.S. import of (and dependence on) foreign oil has soared from 40 percent to over 60 percent. Much of that supply cannot be considered secure. And, huge increases in demand have become the norm for the developing and transitional economies as well as the developed economies. Our ability to effectively address these pressing problems is diminishing daily—and the present oil prices reflect that inadequacy. Yet, some good news is in the realm of the possible—green trading is capable of making a contribution to lessening the negatives around the globe, and through the combined vision of Peter Fusaro and Marion Yuen as well as the supporters of this GreenTrading Summit, the good news is evolving into reality.

SYSTEMIC CHANGES AND INCREASING COMPLEXITY

Since *Energy Decision Making* was written, many changes have made the industry even more complex. At the same time, these changes have underscored the importance of energy—especially electric energy—in the world economy. Let me briefly note several key revolutionary changes: (1) restructuring and open, more competitive markets for energy in a number of economies throughout the world, which have been spurred by growing demand, new players, and initiatives of governmental and multilateral lending institutions; (2) greater dependence on reliable, plentiful electric energy to support a high-tech, information-driven world economy; and (3) globalization of the world energy economy, and the development of trading/futures/options and convergence of financial and physical markets. Yet, many underlying concerns about energy that permeated the 1970s continue unresolved: security, reliability, price stability and accessibility, and resolution of global "energy poverty."

The stakes are higher than ever. The increasing dominance of sophisticated major financial players in controlling positions in worldwide energy markets is a major change. Traders and investment banking houses are acquiring electricity generation and transmission facilities as well as gas production and marketing operations, extending their powerful reach in the international petroleum markets. And these entities are also trading emissions, transmission, and transportation rights. Yes, the petroleum giants, such as ExxonMobil, BP-Amoco, and Shell, continue to be at the pinnacle of the energy world. However, they are increasingly tied to the financial/trading operations as participants, clients, and marketers. The relationship between financial interests and energy interests is intensifying. The creation of regional transmission organizations and open-access pipeline systems also has opened the worlds of trading in supply and transmission or transportation rights, encouraging unaffiliated third-party marketers, arbitrageurs, and aggregators to play ever-increasing roles. These phenomena and the overall demand for capital to build energy infrastructure require an ever-higher level of sophistication. In dealings with financial entities and all who either represent them or clients impacted by them, the stakes are in the billions—not millions—of dollars, and there is escalating pressure to comprehend the complexities of the transactions as well as the laws and regulations that govern them.

Other players that are de facto setting energy policies of the world's nations include major insurance firms and multilateral lending or guarantor institutions concerned about a panoply of risk/reward issues ranging from global climate change to the security of projects in less stable political or regulatory environments. These players demand higher premiums for the insurance they provide to enable the construction of new projects. They also provide environmental safeguards as well as determine the sources and transportation mechanisms of fuel. Experiences from California to India provide cautionary tales to investors, lenders, and guarantors.

The investment community has a heightened concern about the stability of non-financial players in the energy markets. Guarantees, political risk insurance, and other supports are prerequisites, and the expectations are that if the deals go sour, these will be called upon posthaste. The fallout in confidence of both regulators and the financial community in the energy sector has been exacerbated by the Enron debacle and its attendant insolvencies and financial security. Concerns on market manipulation abound and create a "drag" on large energy sector deals, increasing transactional costs.

Global terrorism, including the events of September 11, 2001, also affects perceptions of the energy industry. Currently, the worldwide price of oil is established with reference to a significant "terrorism premium" built into the price of each barrel. Additionally, the worldwide experiences of human error, equipment problems, or force majeure events causing blackouts raise concern for all. The North American blackout of August 2003 has heightened concern about the stability and reliability of the electric infrastructure and has increased scrutiny of legal assumptions behind service obligations and legal consequences of interruptions. The magnitude of the blackout in North America a little less than a year ago makes it particularly notable; it led to a Canada-U.S. comprehensive investigation and report on the event. However, the August 2003 blackout is not an isolated instance; others have occurred and are taking place worldwide. A premium may now be placed on the potential for blackouts in larger deals and in analyzing service obligations of all players in the supply-and-delivery chain. In our increasingly technologically driven society, the disruption of life and economic stability that traditionally happens in a blackout is sharpened dangerously because massive amounts of technological productivity are lost. Simply put, the new reality depends upon reliable and available electric energy for technological and information flow purposes. Yet, ironically congestion of transmission lines, aging plants, and environmental restrictions all impede the realization of this important objective.

Simultaneously, there is an apparent decline in the supply base of oil and gas reserves in many parts of the world.[4] When that assumption appears faulty, the lawyers and regulators descend to deconstruct and criticize the estimation efforts in their totality. Further, force majeure events of various sorts could impact the extraction and delivery of the resources that remain. The resulting uncertainty of reserve estimates naturally shakes investor confidence and piques the interest of financial regulators, further clouding the energy picture.

KYOTO IMPACT—DIRECT AND INDIRECT

The Kyoto Protocol ("Kyoto") will have an undeniable impact on the intertwined energy/environmental systems; whether or not it is fully agreed to by all the major players in the world, it is already an enormously influential consideration. Kyoto is a catalyst forcing companies and countries to take steps to address the greenhouse gas ("GHG") emissions issue in advance of ratification. Whether through carbon sequestration, changes in electric generation and transportation fuel mix with a shift from petroleum or other carbon-based fuels, or the development of alternative energy resource conservation, or green trading and "clear skies" types of initiatives, the plethora of laws and regulatory policies that are emanating from various decisional bodies throughout the world is daunting. Lawyers and regulators ultimately will be responsible for translating the policy shifts into "action documents" and developing laws and regulations to respond to this global concern. The recent decision by Russian President Vladimir Putin that his nation will become a signatory to Kyoto is a major shift, igniting interest and activity in green trading immediately after the announcement. Russia's decision will increase pressure on all countries to participate. Kyoto, which is linked to the much broader world of international trade policy, security, and economic development worldwide, also challenges our legal systems and policy formation. The consideration of the Kyoto issue is further complicated by the present focus on security and antiterrorism concerns. However, it is inevitable that Kyoto looms large. The Bush Administration is once again—post election—promoting a new energy policy with its newly appointed Secretary of Energy at the helm of the initiatives. More significantly, companies, non-profit organizations, and financial institutions are taking steps that lead in the direction of addressing the GHG question.

NEW AND REINVIGORATED SOLUTIONS

On the development side of the energy equation, interest in alternative energy sources as a correlative development to increased technological capability in this important arena is expanding. Increased efficiency and conservation initiatives together with renewables such as wind power, solar energy, hydropower projects, biomass energy, and other new sources will play increasingly important roles in the energy economy. Distributed generation is also providing meaningful solutions.

A basic and important responsibility for lawyers and the multilateral lending institutions is to develop legal frameworks to promote these alternative sources; past experiences must be updated to incorporate the lessons learned from the decades of "legal experimentation." Further development of methanol, electric, or hybrid transport options also will expand the opportunities to address supply needs.

THE OLD BECOMES NEW AGAIN

Natural gas is an essential building block for the international energy economy. Yet, in many parts of the world, it is not geographically close to markets. Thus, the "re-ignition" of other modes of supplying gas demand is taking place. Previously utilized (but somewhat out of favor) approaches to this energy supply dilemma also are re-emerging. The reinvigoration and expansion of the role of liquefied natural gas ("LNG"), particularly in North America, is a remarkable development. Some 30 years ago in the United States, LNG was becoming an attractive alternative to the domestic supply base, which was as constricted then as it is now. However, the interest in LNG declined over the years as the price of oil and natural gas declined, and environmental considerations prevented the siting of many new energy installations. Instability and unpredictability in world energy markets undermined attempts to fully deploy LNG in North America. Times change though; presently in the United States, there are more than 30 projects proposed throughout the country, including in areas where the notion of an LNG terminal would have been unthinkable only five years ago. And existing facilities are expanding rapidly. Of course, there remain impediments to LNG development. Nevertheless, it is believed that a number of major new projects will ultimately be built. At present, the FERC has authorized projects in the Louisiana and the Texas Gulf Coast area; others are likely to move forward in Massachusetts, Georgia, Florida, California, and Northern Mexico. These new projects will be competing for LNG supply in a worldwide marketplace. Additionally, non-conventional natural gas supplies, such as coal bed methane, gas associated with geo-pressurized brine, and other sources are getting another look.

In addition to LNG, nuclear energy is being considered actively once again. Obviously, concerns (with accompanying political gridlock) about disposal of spent nuclear fuel, security of facilities, and security of the nuclear fuel supply itself have prevented new plants from being built in the United States. However, in light of pressures from Kyoto, it is possible that the nuclear industry may experience a renaissance in North America. This is an "unfinished chapter" in the energy policy book.

LACK OF AGREEMENT

Returning to the retrospective, in the United States we have not yet achieved any meaningful form of agreement on key energy issues after years of trying. Since President George W. Bush took office, his administration has been exerting its political will to secure the enactment of a comprehensive energy bill. Repeatedly, that energy bill was delayed from consideration. Nevertheless, it continues to be on the Congressional agenda and remains on the Senate Republican's "Top Ten" list of legislative initiatives. Skeptics are numerous and when any bill will actually result is not clear at all. Even such "non-controversial" provisions such as electricity reliability apparently cannot be enacted. There will be more heated debate on this topic, particularly with the rise of oil prices, or if there are problems on the electricity grid over the summer, or if natural gas supply runs short and prices escalate dramatically, driving up prices beyond their record highs at the moment. Market manipulation and environmental disharmony are certain to further complicate the debate. And after the elections if there is no enactment, the issues will reappear with a new supply of concerns for lawmakers and regulators, and those affected by them.

WHERE ARE WE HEADED?

Today, my questions are these: Is our legal and policy structure keeping pace with the demands created by an increased world population, higher expectations regarding access to energy supply, demand for infrastructure, and the need for reliable and safe energy supply to run a technologically driven society? Have our laws evolved adequately to respond to the environmental and systemic changes around the globe? Have we secured the tools necessary to fully meet the challenge ahead?

---

[1] *Energy Decision Making—The Interaction of Law and Policy,* Joseph P. Tomain and Sheila S. Hollis, D.C. Heath and Company, Lexington, Mass. 1983.

[2] *Ibid.*

[3] *Ibid.*

[4] The word "apparent" may seem curious. However, there are anomalous developments in the reporting of reserves and disconnect between reserve base and actual flow of resources to the marketplace. The full implications of the anomalies between reserve base reporting versus production volumes is yet to be resolved and will be the topic of increasing debate and scrutiny, as it was in the 1970s and 1980s. There exists always the question of the accuracy, completeness, and "shaping" of reserve and production reporting. In the United States, the Securities and Exchange Commission continues to review reporting and analyze the correctness and appropriateness of it.

Sheila Slocum Hollis, is *chair* of the Washington, D.C., office of Duane Morris LLP, and serves on the firm's Executive Committee and Partners' Board. She can be reached at **T**: (202) 776-7810 and **E**: sshollis@duanemorris.com.

CHAPTER 4

Global Greenhouse Gas Markets: Where Do We Go from Here?

By Susanne Haefeli and Einar Telnes

With the Kyoto Protocol ratification having less impact on project developers, investors, insurers, and regulators than initially intended, these parties have been put in a difficult position to decide efficient measures to tackle the global climate-change challenge. The question arises then about how to generate a liquid CO_2 market with great price transparency, many actors, and low entry and exit barriers in this highly uncertain context? This chapter describes the present markets and prices (as of February 2005), and explains how these are influenced and who the main actors are. It then goes on to discuss possible future developments in two of the arenas affecting future markets: the political arena where cultural differences and short-term mandates dictate the agenda and the technological arena where new energy carriers might ease the great task of mitigating climate change.

PRESENT MARKETS

At present, Greenhouse Gas (GHG) credit markets are still relatively small, scattered, and illiquid. Reasons for this low level of development are mainly related to the lack of political will to implement measures that consistently reduce anthropogenic GHG emissions. This current situation is driven by the complex scientific evidence of climate change, the effectiveness of possible mitigation measures, and the perceived impact on the economy. Therefore, the present demand for emission reductions is limited, which has not encouraged further establishment of the emission reduction market.

Despite the present limited and scattered market, some clear trends are pointing toward a more firm and consistent future developmental path: The most mature and dynamic market is currently emerging under the E.U.

Emissions Trading Scheme (E.U.-ETS) which started on January 1, 2005. This will have a significant impact on all major emitters of CO_2 in Europe, and will be extended to other GHG emissions in 2008. The E.U. scheme has established links toward the Clean Development (CDM) and Joint Implementation (JI) mechanisms of the Kyoto Protocol. These links are likely to contribute to a higher demand for GHG credits within the E.U. and, in turn, provide for further liquidity and expansion of the total market. However, the demand for emission reduction credits in the E.U.-ETS will depend on the amount of allowances each country distributes among its national trading participants. A firm cap for the companies that represents the Kyoto commitment and E.U. burden sharing agreement has successfully been enforced by the European commission. This will leave little chance of E.U. allowance oversupply and a greater opportunity for increased demand of CDM credits. The E.U.-ETS is also in the long term likely to be linked to emission trading initiatives in Canada, Norway, and Japan. Also individual Australian states are investigating how they can link their emission trading systems to the E.U.-ETS.

Emission reduction plans have also cropped up in some individual states of the United States. Despite GHG emissions caps' very low priority on the federal level, due to the perceived poor impact on the economy, "the cap and trade market" is expected in California, Washington, Oregon, and some of the New England states, which are planning or have already passed legislation that will place limits on GHG emissions. While this is on a voluntary basis in some places, some states and sectors are also legislating mandatory requirements. Although these programs are in their infancy and partly voluntary, it is possible that they will converge and form the basis for a federal system in the United States in the long run.

In addition to the regulated operations described above, a non-regulated part of a GHG market has also emerged. Companies have for the past six to eight years traded verified emission reductions (VERs) as a voluntary way to offset the carbon footprint from their operations. Although with a high uncertainty on price and compliance acceptance, these actions have contributed to the development of the voluntary programs now being put in place, such as by the World Economic Forum's GHG registry and the Chicago Climate Exchange program.

These programs' feasibility depends on achieving more transparency and lower prices, which, in turn, is dependent on the willingness to compromise on a common set of rules. The rules are likely to have a strong binding on international treaties such as the Kyoto Protocol. In addition, agreement is needed on how technical issues such as emission baselines are to be accounted for and verified, and whether and how emission intensities can have a role in a market that aims to reduce the overall concentration of GHGs in the atmosphere. The International Standards Organisation (ISO), World Business Council for Sustainable Development (WBCSD), American Petroleum Institute (API), and other institutions are currently developing standards and guidelines that aim to contribute to this overall objective. Short-term prospects for rule convergence are not good, though, due to limited cooperation and scope overlap represented by these initiatives and the reluctance of the United States and Australia to commit to any absolute targets. However, present political signals point in the direction that these countries will have to commit to reductions in the long run. Hence, a limited optimism for long-term future viability of a strong and consistent market can be exercised.

From being a rare feature for this market, options for CERs under the Clean Development Mechanism and the EU and UK Emissions Trading schemes are now traded frequently. Although these trades do not yet represent big volumes, they show a clear anticipation from market actors on the need to hedge their GHG positions. This activity is also signalling a change among the market actors: from being a market dominated by the "altruists" (such as the World Bank's Prototype Carbon Fund, the Dutch Government and early corporate movers), the buyers of carbon credits now seem to divide in four main fractions:

1. Individual countries with need for GHG credits to meet Kyoto commitments. Among these are Denmark, Finland, Italy, The Netherlands, Sweden, Spain, and Austria.
2. Individual businesses with need for credits to meet caps imposed by the E.U.-ETS. Among these are the main oil companies, cement manufacturers, and the large European power companies.
3. Banks and insurance companies engaging in the GHG market through building of funds with differentiated GHG project portfolios. Some of the large German banks and Swiss insurers have programs under development for such funds, in addition to the Japanese Bank for Industrial Cooperation (JBIC).
4. In addition to the groups mentioned above, there are also speculators entering the market. In anticipation of increasing future carbon values, some companies invest in projects and take positions in the market in order to gain financially from this in the long run. Such

speculators are represented by project developers as well as consulting companies with GHG interests.

Knowing that early movers may have advantages in picking good and inexpensive projects (but also disadvantages related to risks and transaction costs), no one can wonder why many companies are interested in taking early GHG positions. On the supply side over 300 projects are currently presented under both the CDM and the JI scheme, ranging from 20,000 to 1,000,000 tCO_2 equivalents annually. These projects predominantly originate in the renewable electricity generation and the landfill gas sectors with some initiatives also taking place for energy efficiency and fuel switch measures. At present, most CDM projects are in Southeast Asia and Latin America, and nearly half of the latter are located in Brazil.

PRESENT GHG CREDIT PRICES

During the last years, the initial high prices quoted and predicted for GHG credits in 1998–99 at US$30 per tCO_2 have come down to a significantly lower level. The Euro has also partly displaced the U.S. dollar in GHG price indications and quotes due to the dollar's depreciation. Distinct price differentiation can in addition be observed between so-called "compliance credits" under the E.U.-ETS and Kyoto Protocol credits with prices ranging from 9 euros for E.U. allowances down to 3 euros for some CERs. This difference is most probably going to stay until the second half of 2005. Then it is expected that the E.U.-ETS will force rapid price discovery with the "ceiling" of the "price house" defined by the marginal abatement costs of individual companies and the "price floor" being determined by the generation cost of credits from external projects (CDM, JI). When looking at the recently published national allocation plans for E.U. allowances, the trend to allocate allowances rather generously suggests a price development further towards the CERs/ERUs "floor." Whether this will remain in the midterm is uncertain.

Several observations can be made from this picture:

- The low prices for CERs/ERUs suggest that there are enough projects identified that can generate cheap GHG credits. However, the CDM/JI is not a one-fix-for-all solution in mitigating climate change and contributing towards sustainable development. A number of sound projects are not implemented due to high transaction costs

related to project baseline study, delays in approvals from government agencies, regulatory uncertainties, etc.
- The danger of a segmented market and differentiated prices is still not barred as government and private actors will have different selection criteria when buying emission reductions; for example, government transactions may be guided more by preferred trade partners and a wide range of non-economic criteria.
- In the effort to avoid double counting in the E.U. zone, ERUs from JI projects will likely be generated predominantly in the renewable and waste sectors, as combustion units in the new E.U. countries fall under the E.U. Emissions Trading Scheme.

In the long run, a convergence between the different "currencies" such as AAUs, ERUs, CERs, EUAs, in the GHG market is likely to develop because market actors will have identified their marginal carbon abatement costs and will have a need for fungibility between the assets that can be used to offset their emissions or meeting their compliance targets. A lack of fungibility between these currencies can be seen as a distortion to a market that is supposed to contribute to an optimal solution to a global problem.

POSSIBLE FUTURE SCENARIOS

Climate change is not going away with the Kyoto Protocol. If the objective of atmospheric CO_2 concentrations less than 550 ppm should be reached, the Kyoto Protocol will represent only the start of many attempts to limit human induced climate change. As a long-term problem, climate change will also need long-term solutions that reach far beyond the first Kyoto Protocol commitment period, which ends in 2012. Presently, the Kyoto Protocol is still the main market driver with support of OECD countries except for the USA and Australia and most of the developing countries, such as China, India, and Brazil. Future scenarios will depend on the development in the two following arenas:

Political Arena

The E.U. is likely to lead further international GHG reduction efforts, supported by Japan and Canada. Although the present initiatives to tackle emission reductions in the United States and Australia so far are driven from individual states rather than the federal governments, we can assume that these initiatives in some way will link up with the Kyoto Protocol in the long run.

In order to understand the current discussions on beyond-Kyoto scenarios, it is useful to point out some of the approaches and inherent weaknesses of the Kyoto Protocol:

There is no clear principle or logic behind the differentiation of commitments, i.e., how Annex I countries got their targets. The division between developed and developing countries will be problematic in future climate change negotiations, as developing countries become more advanced and also contribute more to global emissions. Furthermore, the negotiations in Kyoto resulted in an agreement on emission targets, but left the "rules of the game" unsettled. The post-Kyoto negotiations on these more technical issues have opened the door for renegotiating the commitments. It might be wiser in the future to try and first reach agreements on the rules of the game before subsequently determining the commitments.

Renegotiations on an international level beyond the first commitment period of the Kyoto Protocol is likely to involve areas beyond project-based mechanisms and emissions trading. Two of the organizations with members representing a significant share of global transport emissions, the International Maritime Organisation (IMO) and International Air Traffic Association (IATA), are already looking into how these emissions can be reduced. In addition, products that result in less GHG emissions on a life-cycle basis than competing products may be eligible for "product-related" or "value-chain" credits. Different incentives can be put in place to promote such products, as we already see under the Australian Greenhouse Friendly Programme, a program where products sold are made CO_2-neutral by a link to emissions-reduction projects.

The Technology Arena: Can Renewable Energy Sources Deliver?

The International Energy Agency's (IEA) World Energy Investment Outlook (WEIO) states that US$16 Trillion would be needed in order to provide access to basic energy needs for the entire world population (IEA, 2003, *World Energy Investment Outlook*, Paris). The WEIO furthermore clarifies the magnitude of the challenge with separate prospects, costs and supply contexts for each known fuel. The predominance of fossil fuels, especially coal, is not going away for the next thirty years—partly due to the present growth in China and India. In turn, renewable energy sources need faster implementation and alternative energy carriers such as hydrogen need more urgent development in order to limit the emission growth for the next decades. This must be done at affordable energy prices that are acceptable to all of the world's population, not only the rich ones.

Hydrogen is not likely to dominate the climate-change solution for a decade or two. With exceptions, the technology is still costly due to the energy penalty and continues to generate GHG emissions during processing. In addition, the necessary supply infrastructure requires large investments. Last but not least, the political willingness to support hydrogen as the key solution is limited, which means that a sustainable and cost-effective solution is still far away. What if the USA had put in place commitment and programs for hydrogen as the main energy carrier of this century equal to the U.S. moon-landing program of the 1960s? In that case, we would be able to look back in the future and see that the solution was much closer that we thought in 2005.

The energy intensity objectives put forward by the current U.S. administration will not matter much in limiting global GHG emissions. As long as the net emissions from the world's largest GHG emitter continues to grow, neither energy intensity objectives nor voluntary company targets will contribute much in the global picture. In order to advance economical development, the developing countries will have to increase their emissions. Yet, they will have no political will to curb emissions as long as the industrialized countries continue to increase their net emissions. For developing countries, renewable energy sources like agricultural waste—but also wind, solar and tidal power and small hydropower installations—are likely to play a significant role in the growth of energy supplies. However, as many developing countries have a tremendous economic growth, they will also be dependent on fossil-based energy supply as long as other cost-effective solutions are not available. In this context, it is amazing to see that wind-powered turbines that four years ago were not competitive against conventional power generation, now are installed and operated at close to competitive prices due to economies of scale in the manufacturing of wind turbines.

Seen in the global context, this also means that the CDM is the only instrument that involves all of the countries where more than two thirds of the energy investments need to take place. This is a very clear justification for the CDM to remain in place over time, and for institutional capacity to be developed in order to sustain the use of the mechanism.

FUTURE PROSPECTS

From above discussions, the following can be concluded. The world will evidently face a carbon-constrained economy. It is just a matter of time before the inequity in emissions per capita between the developed and developing world will cause more firm measures to be put in place to balance out, and in a direction that is more acceptable to the poor countries of the world. It is our opinion that the successful implementation of the Kyoto Protocol will be an important milestone in this regard, and a clear signal that market-based mechanisms are a way to achieve emission-reduction objectives.

Besides this, there is also a need for non-Kyoto states and different GHG registries to align their schemes and related requirements in order to provide for market liquidity and, in the long-term, aim for complete market fungibility. In the period beyond 2012, credit prices are likely to increase, as countries and companies will have even more strict targets than those being agreed under the Kyoto Protocol and in the E.U.-ETS. Given these increased constraints, the demand for credits are likely to rise further, thus providing for multiplication of the current market size and real market liquidity. This again will spur off a number of more innovative approaches on how to limit and reduce global greenhouse gas emissions.

Susanne Haefeli is *project manager* and **Einar Telnes** is *technical director* of DNV Certification's international climate change services. Mr. Telnes can be reached at **T:** 47-913-52-877 and **E:** Einar.Telnes@dnv.com.

CHAPTER 5

Initial Observations from the First Year of the Chicago Climate Exchange

By Richard L. Sandor and Claire M. Jahns

September 30, 2004, marked the one-year anniversary of the Chicago Climate Exchange's first auction of Carbon Financial Instruments. The discriminating price auction provided the first market price information for carbon-dioxide allowances within the framework of a multisector, multinational, and legally binding emissions-reduction program. The 2003 auction followed extensive design and preparation phases of the Chicago Climate Exchange (CCX) and the beginning of greenhouse gas (GHG) emissions market activity. Perhaps the anniversary of that day is a good opportunity to recap the activities of the exchange to date, with particular emphasis on the events and milestones of the last 12 months. This account will also touch on CCX's plans for the future, which include an Amsterdam-based subsidiary company to engage the European Union Emissions Trading Scheme as well as futures and cash contracts in other environmental commodities in North America.

CCX HISTORY AND OBJECTIVES

Before reviewing CCX's developments since September 2003, it is valuable to state the motivation behind the establishment of CCX. The overarching goal of CCX is to turn the theory of greenhouse gas (GHG) emissions reduction and trading into a reality—to prove that the concept can work in practice. Nine formal objectives of CCX were articulated during the design phase:

- Demonstrate unambiguously that a cross-section of North American companies, schools, municipalities, and other entities can reach agreement on a voluntary commitment to reduce greenhouse gas emissions and implement a market-based emissions-reduction program;

- Establish proof of concept by demonstrating the viability of a multisector and multinational greenhouse gas emissions cap-and-trade program supplemented by project-based emission offsets;
- Create a mechanism for achieving price discovery as well as developing and disseminating market information;
- Allow flexibility in the methods, location, and timing of emissions reductions so that greenhouse gas emissions can be reduced cost-effectively;
- Facilitate trading with low transaction costs;
- Build market institutions and infrastructure as well as develop human capital in greenhouse-gas emissions trading;
- Encourage improved emissions management;
- Harmonize and integrate with other international or sovereign trading regimes; and
- Develop a market architecture that rewards innovative technology and management as well as encourages sustainable farming and forestry practices.

We will begin with a brief overview of the design phase of the CCX program, which led up to the September 30, 2003, auction. The CCX design phase, funded by two grants from the Chicago-based Joyce Foundation, ran from 2000 until the publication of the Chicago Accord in August of 2002. More than 50 participants from the industrial, electric power, oil and gas, and forest products sectors—together with offset providers, a variety of service providers and municipalities—participated in the design phase of what was originally to be a regional emissions trading program in the Midwest. By the end of the design phase, however, it had become apparent that the concept of GHG emissions trading would be better tested in a broader North American framework. Moreover, the design-phase participants recognized that a multisector, multinational framework would better prepare them for possible mandatory emissions trading programs.

CCX assembled a group of public directors between the end of the design phase and the 2003 auction. Each of the public directors brings high-level and unique professional skills and perspectives to the CCX board of directors. The public directors serve alongside three CCX Member officers, the CCX Chairman, and the CCX President on the Board of Directors, and are listed in Table 1.

Table 1
Public Directors (2004) of the Chicago Climate Exchange

Les Rosenthal (Vice-Chairman) Principal, Rosenthal Collins; former chairman, Chicago Board of Trade
Maurice Strong (Vice-Chairman) Chairman, Earth Council; former undersecretary-general of the United Nations
Warren Batts Professor, University of Chicago Graduate School of Business; former CEO of Tupperware
Governor James R. Thompson Chairman, Winston & Strawn; former four-term governor of Illinois
Governor Christine Todd Whitman Former U.S. EPA administrator; former governor of New Jersey

The Chicago Accord is a summary of the CCX market architecture developed in the design phase. The CCX Rulebook, which was released a year later in August of 2003, fleshes out the framework put forth in the Chicago Accord and provides detailed protocols for measuring, reporting, transacting, and retiring of Carbon Financial Instruments (CFIs), the commodity traded on CCX. The Rulebook also sets forth rules relating to exchange governance, verification of GHG emissions and offsets as well as other matters necessary for the orderly functioning of the program. The CCX Rulebook is the first articulation of a comprehensive market structure and governance system for a multisector and multinational GHG trading program. The exchange received extensive input from its members, as well as consultants, independent counsel, and the National Association of Securities Dealers (NASD, which is CCX's provider of regulatory services) in drafting the Rulebook. By working with these rules on an operational basis, CCX members integrate quantifying, reducing, and trading GHG emissions into their day-to-day activities.

It is important at this point to reemphasize that CCX is a pilot program. While the Rulebook sets the standards for CCX, the viability, stringency,

Table 2

Key Features of the Chicago Climate Exchange

Geographic Coverage	GHG emission sources in the United States, Canada, and Mexico, and offset projects in North America and Brazil
Emission Targets and Timetable	Legally binding emission reduction commitments for the years 2003 through 2006. Emission targets are 1% below baseline during 2003, 2% below baseline during 2004, 3% below baseline during 2005, and 4% below baseline during 2006.
Emission Baseline	Average of annual emissions during years 1998 through 2001
Gases Included	CO_2, CH_4, N_2O, PFCs, HFCs, SF_6
Emission Offsets	Sequestration of CO_2 in North American soil and forest biomass projects and methane destruction at North American landfill and agricultural sites; additional projects eligible in Brazil.
Early Action Credits	Credits from specified early projects to be included starting in 2004
Registry, Electronic Trading Platform	Registry will serve as official holder and transfer mechanism, and is linked with the electronic trading platform on which all trades occur.
Exchange Governance	Self-regulatory organization overseen by committees comprised of exchange members, directors, and staff
Renewable Energy and Environmental Innovation	Rewarded in design
Verification and Regulatory Services	Provided by NASD, leading private sector provider of financial regulatory services

and sufficiency of the protocols it contains are put to the test in day-to-day operations. The fundamentals of the program—the 1998 through 2001 baseline period, the 1 percent per year below-baseline emissions reductions required of members, the verification requirements, etc.—will remain unchanged until the end of the pilot in 2006. The rules do, however, allow flexibility in the methods, location, and timing of reductions so that GHG emissions can be reduced cost-effectively. This flexibility also gives participants the opportunity to compare and contrast the different ways in which they track emissions internally and identify least-cost emissions reductions. A major value proposition of CCX is that participants can add to their own experience by learning from their fellow members. Members' experiences in meeting the fundamental requirements of the program are discussed formally—in committee meetings and other member gatherings—and informally, in casual phone conversations or at other gatherings such as industry events.

FIRST-YEAR DEVELOPMENTS

We can now consider some first-year indications of the achievements of the CCX program. The following sections will discuss program participation rates, GHG emissions submissions and audits, environmental progress and trading activity.

Membership

The exchange now includes more than 70 members from a dozen sectors of the economy, academia, and the public sector. See Table 3 for a list of members as of October 2004. Florida-based TECO Energy, Inc., joined CCX in August of 2004, and Vermont-based Green Mountain Power joined in early October. The past year also saw the addition of industrial energy users such as IBM and Rolls-Royce, as well as the University of Iowa and the University of Oklahoma. The aggregate direct GHG emissions of CCX members are in excess of 200 million metric tons of carbon dioxide, and this would make the CCX membership one of the largest "countries" in the European Union's mandatory carbon dioxide market, which will launch in 2005.

A new membership designation—Associate Membership—was created in the early spring of 2004 to meet the desire of numerous low-emitting entities to learn the practical details of carbon trading. Associate Members

Table 3

Members of the Chicago Climate Exchange (October 2004)

Chicago Climate Exchange® Members

* October 2004 *

Aerospace and Equipment
Rolls-Royce

Automotive
Ford Motor Co.

Chemicals
Dow Corning
DuPont

Consulting
Domani LLC
Global Change Associates
Natural Capitalism, Inc.
Rocky Mountain Institute

Diversified Manufacturing
Bayer Corporation

Electric Power Generation
American Electric Power
Green Mountain Power
Manitoba Hydro
TECO Energy Inc.

Electronics
Motorola, Inc.

Energy Management Services
Sieben Energy Associates

Environmental Services
Waste Management, Inc.

Food Processing
Premium Standard Farms

Forest Products
International Paper
MeadWestvaco Corp.
Stora Enso North America
Temple-Inland Inc.

Information Technology
IBM
Open Finance LLC

Legal Services
Foley & Lardner

Liquidity Providers
AGS Specialists LLC
Amerex Power Ltd.
Michael R. Anderson
Raymond S. Cahnman
Calyon Financial Inc.
Thomas H. Dittmer
Eagle Market Makers, Inc.
Evolution Markets LLC
FCT Europe Limited
First New York Securities LLC
Goldenberg, Hehmeyer & Co.
ICAP Energy LLC
Chris J. Johnson
Kingstree Trading LLC
Kottke Associates LLC
Marquette Partners LP
Glenn M. Miller
Douglas M. Monieson
Natsource LLC
Rand Financial Services, Inc.
Refco LLC
Serrino Trading Co., Inc.
Shatkin Arbor, Inc.
C. Richard Stark, Jr.
Jeffrey B. Stern
Lee B. Stern
The League Corporation
Tradelink LLC
Tradition Financial Services
Transmarket Group LLC

Municipalities
City of Chicago

Non-Governmental Organization
American Coal Ash Association
American Council on Renewable Energy
Houston Advanced Research Center
World Resources Institute

Offset Aggregator
Iowa Farm Bureau

Offset Provider
Klabin S.A.
Restoration Soil & Research, Ltd.

Pharmaceuticals
Baxter Healthcare Corporation

Private University
Tufts University

Public University
The University of Oklahoma
The University of Iowa

Religious Organizations
The Jesuit Community of Santa Clara University

Semiconductors
STMicroelectronics

Steel
Roanoke Electric Steel Corp.

Technology
Ecoenergetics srl
Millennium Cell

Transportation
Amtrak

such as the World Resources Institute commit to fully offset their indirect emissions from business-related energy consumption and business travel by purchasing and retiring CCX CFIs. Like CCX members with direct emissions, Associate Members voluntarily adopt this commitment for the four years of the pilot program, 2003–2006. New Associate Members include consulting companies such as the Rocky Mountain Institute, Natural Capitalism, Inc., and Global Change Associates; legal firms such as Foley & Lardner; and nongovernmental organizations such as the Houston Advanced Research Center.

The year 2004 also saw a significant increase in the number of liquidity providers participating in the CCX market. More than 20 proprietary trading firms, emissions trading firms, and individual traders have joined CCX. The liquidity provided by these traders is critical to the development of a successful market, and CCX is encouraged by the heightened interest from the trading community. New CCX endeavors, which are explained in the final section, are expected to further engage proprietary and institutional traders.

The first annual CCX membership meeting was held June 16–18, 2004, in Chicago. More than 70 individuals participated, including representatives from 25 CCX member companies. The program consisted of presentations by CCX Chairman and CEO Richard L. Sandor and CCX Senior Vice President Michael J. Walsh. Several CCX committees held meetings while other participants received training on use of the electronic trading platform and met in informal groups to discuss the program. The annual meeting was an opportunity for members to share their experiences with CCX to date as well as to begin reviewing lessons learned and developing future plans.

Emission Baselines and Audits

The members of CCX have undertaken significant efforts to assemble internal records of energy and materials use. For members in the electric generation sector, GHG emission reports are based on the data from the Continuous Emission Monitors (CEMs) installed on most electricity generating units. Where CEM data are not available and for most members in the industrial and manufacturing sectors, the GHG emissions are determined by applying emission factors to fuel inputs. CCX rules prescribe the use of World Resources Institute/World Business Council for Sustainable Development (WRI/WBCSD) calculation tools and other methods to determine GHG emissions from fuel use. For several unique industrial processes, CCX worked with experts to develop emission factors, as none had existed prior.

All emissions baselines were audited by NASD, which was also contracted to perform an analysis of each member's GHG emissions data for the 1998–2001 baseline period and for each year of the pilot program, 2003–2006. The NASD analysis, which was based on CCX rules and procedures, is designed to determine whether emissions data submitted is congruent with actual emissions claimed by the member. This is based on a review of actual proofs of energy usage or purchase, such as fuel purchase receipts or electricity bills. These audits often identified issues needing resolution by a member or suggested various needed modifications to CCX rules.

CCX and NASD developed the auditing process after thoroughly researching into existing emissions trading programs and calculations developed by WRI and the Intergovernmental Panel on Climate Change (IPCC). Each member's review is tailored so that the requisite data can be obtained for review of compliance with CCX rules without subjecting the member to an undue imposition of time and resources. Once an initial review of each member's baseline has been completed, NASD determines a sample size from among the baseline submissions to be analyzed for compliance with the rules of the exchange. NASD then issues a request for third-party documents ("proofs") to support the sampled baseline elements. After the member provides the proofs, they are compared to the sample data and checked for accuracy and completeness. Upon completion of the analysis, NASD provides feedback to the member identifying specific discrepancies or errors in the original baseline or 2003 emissions report. This communication allows the member to address areas of concern identified by NASD staff and, where appropriate, to amend the reports. The NASD's review process contributes to the credibility of the CCX program and, according to some members, to the value of their participation. The NASD provides each member with a GHG emissions history verified by the same regulatory body that the U.S. Congress has designated to oversee virtually every public securities firm in the country. In a carbon-constrained world, this verified and peer-reviewed emissions history could give CCX members an advantage over their competitors.

Spring of 2004 marked the first conditional CCX "True-up." True-up is the annual surrender of CFIs by each CCX member in an amount sufficient to fulfill its CCX commitments and is viewed by some as the "moment of truth" for the program. The true-up of 2003 emissions required NASD to intensify its activities in relation to the verification of members' baseline and 2003 emissions. During the 2003 true-up process, each member with direct emissions compared its actual 2003 GHG emissions to the 2003 vintage CFIs held in its registry account. If actual emissions were above the holdings in the registry, that member had to purchase CFIs to satisfy its emission reduction commitment. If actual 2003 emissions were "on target" with the member's emission reduction schedule, that member was deemed to be in conditional compliance and had no need to purchase additional CFIs.

All CCX members with direct emissions achieved compliance with their 2003 emission reduction targets. In fact, members with direct emissions, as a group, had 2003 aggregate emissions 8 percent beyond the 1 percent-below-baseline required by the CCX emission reduction schedule. The annual true-up process is overseen by the CCX members that serve on the CCX Environmental Compliance Committee.

Trading

Shortly after the September 2003 auction, CCX began beta testing its electronic trading platform and online Registry. CCX contracted the IntercontinentalExchange™ (ICE™) to provide, design, and service CCX's electronic trading platform. Atlanta-based ICE is the leading electronic venue for the trading of over 600 energy and metals commodities, and is accessed by more than 5,000 users daily. ICE's existing clientele of energy traders was among the reasons CCX chose to white-label the company's software. As with many other aspects of the CCX program, the designers strove to make the links between this new emissions market and existing, related markets (i.e., electricity, fuels, etc.) as straightforward as possible. The trading platform is relatively simple, with columns for bid and ask prices and quantities as well as a ticker at the bottom. It was designed to be used by everyone from the first-time participant in a commodities market to the most experienced professional energy trader. Each CCX member's online registry page is essentially a bank account of that member's holding of CFIs in each of the vintage years, representing the four compliance years of the CCX program: 2003, 2004, 2005, and 2006. The registry pages also display each member's transaction history. The trading platform and registry pages are Web-accessible through a secure site.

Figure 1

The First Day of Trading on CCX

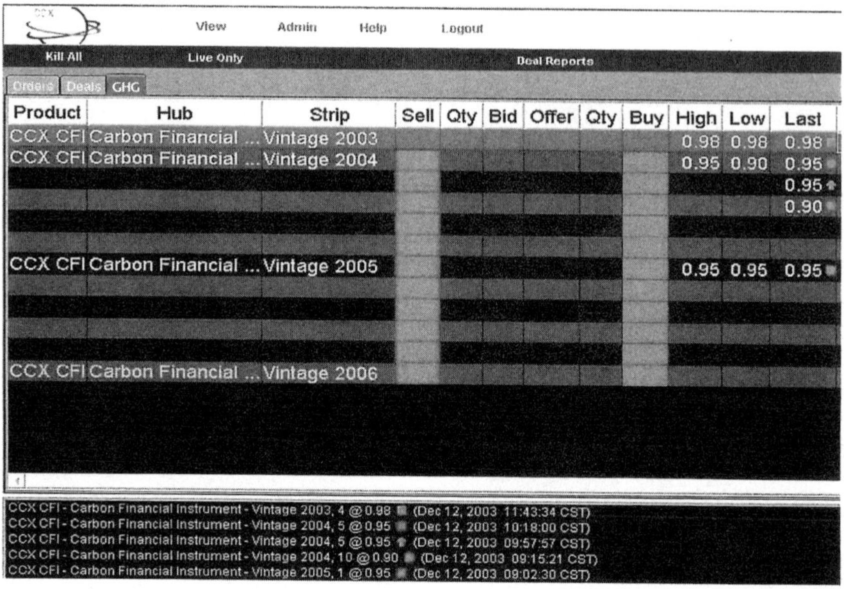

Continuous trading of CFIs, each of which is equal to 100 metric tons of carbon dioxide equivalent, began on December 12, 2003. Figure 1 is an image of the CCX electronic trading platform at the end of the first day of trading. Members reached the milestone of one million tons traded by July 1, 2003. The first nine months of 2004 saw aggregate trading volume of 1,383,000 metric tons' worth of CFIs. September 2004 had the second-highest monthly volume to date, and trading looks to be accelerated in early October. Figure 2 is an image of the live CCX electronic trading platform taken on October 4, 2004. Figures 3, 4, and 5 show summary trading statistics through early October, monthly trading volumes through September 2004, and 2004 vintage CFI prices through early October.

Figure 2

Trading on October 4, 2004

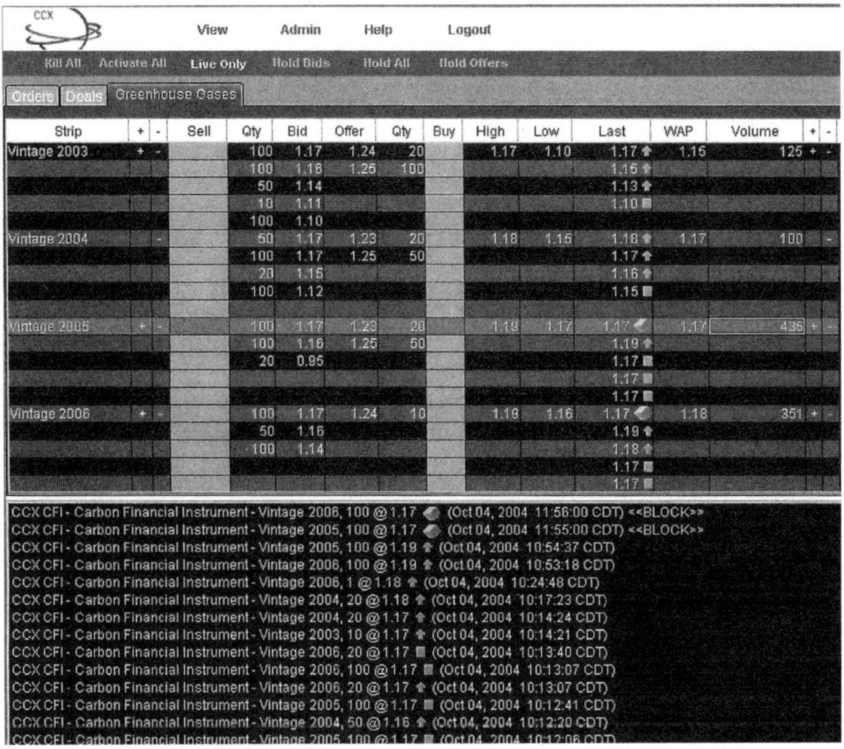

Figure 3

CFI Summary Trading Statistics (December 12, 2003–October 4, 2004)

	Prices ($/metric ton)				Volumes (metric tons)	
Vintage	**High**	**Low**	**Close**	**Number of Trades**	**Traded Volume**	**Total Volume***
2003	1.17	0.73	1.17	104	127,100	227,100
2004	1.18	0.77	1.18	153	469,200	469,200
2005	1.19	0.71	1.17	106	546,800	571,800
2006	1.19	0.80	1.17	78	385,900	385,900
				441	**1,529,000**	**1,654,000**

Figure 4

CCX Monthly Trading Volumes (December 2003–September 2004)

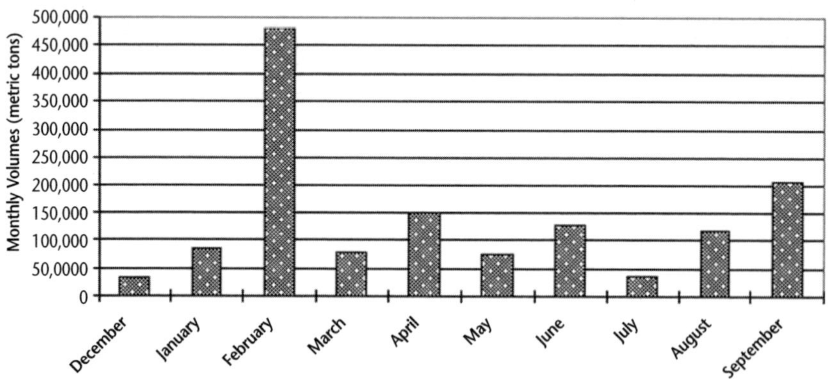

Figure 5

2004 Vintage CFI Prices (December 12, 2003–October 4, 2004)

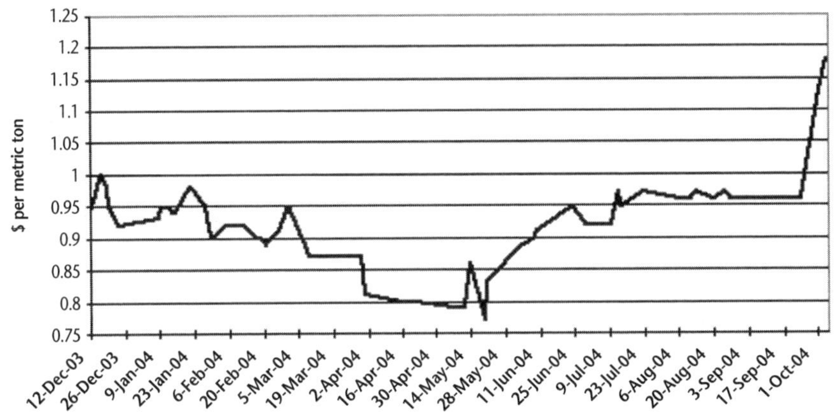

LOOKING AHEAD

The activities of the past twelve months have made 2004 a milestone year for the Chicago Climate Exchange. By undertaking the significant effort required to reach their commitments, the CCX members have begun to prove that a diverse cross-section of North American companies, municipalities, and other entities can turn the concept of exchange-based greenhouse gas emissions trading into a reality. CCX and its membership continue to refine the North American GHG program, as well as extend the concept of allowance trading to other environmental commodities and into other geographic regions.

Other Environmental Commodities

CCX intends to extend the central exchange model not only to non-GHG air pollutants but also to other environmental commodities. For example, recent price volatility in the over-the-counter U.S. sulfur dioxide market suggests that both compliance traders (i.e., U.S. electric utilities) and speculative traders would benefit from increased transparency, liquidity, and clearing services. In addition, CCX is currently engaged in research that explores the feasibility of using market-based mechanisms to address issues of both water quantity and quality. Clean water is becoming increasingly scarce in large portions of North America and in other areas around the globe. CCX is exploring the practicality of establishing markets that would foster more efficient use and stewardship of these and other natural resources.

Chicago Climate Futures Exchange

CCX announced the creation of its futures subsidiary, the Chicago Climate Futures Exchange® (CCFE®), at a futures industry conference held in Burgenstock, Switzerland, on September 9, 2004. As of the date of this publication, CCFE has applied for "designated contract market" (DCM) status with the Commodity Futures Trading Commission (CFTC). CCX has entered into an agreement with the National Futures Association (NFA) to serve as the regulatory services provider for CCFE. The Clearing Corporation will provide financial guarantees and clearing services for CCFE. ICE will continue to provide and service CCX's electronic trading platform for both the cash market and the futures market. The first product to be offered on CCFE will be sulfur dioxide futures, which are expected to commence trading by December 6, 2004.

European Climate Exchange

Finally, CCX is extending its presence to Europe with the creation of the European Climate Exchange® (ECX®). ECX products will provide a central, cleared exchange platform for trading of carbon dioxide emission allowances issued by EU member states under the EU Emissions Trading Scheme (EU ETS). ECX is a wholly owned subsidiary of CCX, and will serve as a sales and marketing office for European products. Based in Amsterdam, ECX is led by Peter Koster, former CEO of Fortis Bank UK and a former non-executive director of the London International Financial Futures Exchange (LIFFE). On September 7, 2004, CCX and the London-based International Petroleum Exchange (IPE) signed a formal Cooperation and Licensing Agreement for the listing of cash and futures contracts in ECX Carbon Financial Instruments (ECX CFI®s) on the IPE. ECX plans to launch futures contracts by the end of 2004, in anticipation of the start of trading under the European Union Emissions Trading Scheme (EU ETS) in January of 2005. Cash products will be launched in early 2005.

The ECX CFI futures contracts traded on IPE will provide participants in the EU ETS with low-cost, financially guaranteed tools for managing their exposure to price volatility in the emerging emission allowance market. The significant fluctuation in the forward price of EU carbon dioxide emission allowances observed to date is a reminder that price movements in this innovative new program introduce significant risks to regulated entities in the EU ETS. Prudence suggests that those exposed to this risk are well-served by assessing their tolerance for risk exposure and formulating risk management and trading strategies that address the exposure.

INITIAL OBSERVATIONS

CCX has always been envisioned as a multisector program that helps build institutions and skills while demonstrating the feasibility of a GHG emissions cap-and-trade program supplemented by project-based carbon dioxide offsets. CCX's central features meet the stated goals of the electric utility, industrial, and other design-phase participants and also represent a credible first step to addressing a complex challenge. The pilot nature of the program has allowed participants to test rules and procedures in a cooperative environment, while also achieving measurable and verifiable GHG emission reductions.

Initial observations confirm that CCX members are rapidly growing their understanding of market-based carbon emissions management through hands-on experience. The GHG emissions data collection process has been easier for some members than for others, yet all members can learn from the obstacles faced by a few. In addition, participants' experiences during this first year of market operation suggest that what appear to be relatively simple requirements can become complicated when put into practice. Through participation in CCX, however, members are able to develop best practices that allow the activities of emissions data gathering, auditing, and trading to become simplified and standardized.

Rigorous emission audit reports by NASD provide an asset that members are presenting to shareholders, the public, and other greenhouse gas initiatives. Important details such as data management and trading procedures, and critical policy issues such as baselines, liability limits, and eligible offset projects, are being understood by a diverse participant group. As an end-to-end program in which members go from initial emissions calculations to an audit process to the trading screen and finally to a spring true-up, CCX has begun to test nearly every aspect of a functional GHG emissions cap-and-trade system. This is ideal for a pilot program whose intention is to prove that the concept of GHG emissions trading can be turned into an economically efficient and environmentally effective practice.

CCX looks forward to the continued expansion of its membership, CFI trading volume, and new cash and futures products. For more information on the Chicago Climate Exchange, please visit the website at www.chicagoclimateexchange.com.

Richard L. Sandor is *chairman* and *CEO* of the Chicago Climate Exchange, and **Claire M. Jahns** is *associate economist* at the Chicago Climate Exchange. They can be reached at T: (312) 554-3350, and E: rsandor@chicagoclimateexchange.com and cjahns@chicagoclimateexchange.com (respectively).

CHAPTER 6

White, Green, and Black Certificate Trading

The Italian Experience

By Stefano Alaimo

Italy has demonstrated a growing sensitivity to environmental issues in recent years. The stabilization of GHG [Greenhouse Gas] emissions has become an important goal in national politics, with many measures adopted to promote economic sustainability in electricity production with renewable sources and energy end-use efficiency.

Since 1992, electricity production from renewable sources has been stimulated by Resolution 6/92 of the Interministerial Price Committee ("CIP 6/92"), which introduced a feed-in tariff for renewable and assimilated sources. As the cost of electricity production can vary, producers receive a different value of incentives for each source used. Recently the Green Certificate market mechanism [hereinafter "GC"] has been introduced, and it will completely replace the administrative price for renewables within some years.

Beside those measures, two energy-efficiency decrees approved in 2001 and recently modified, requiring electricity and gas distributors to achieve target savings on their customers' total consumption. The mechanism also introduces an Energy Efficiency Certificate Market, which is slated to start during 2005.

The ratification of the European Directive 87/2003 by the Italian government also introduces black certificate trading in Italy as well as other E.U. member states. Also, this market will start during 2005. Black certificates are international and relate to GHG reductions.

GREEN CERTIFICATE MECHANISM

The Green Certificate Mechanism introduced a market mechanism in Italy to stimulate electricity production from renewable sources, with the goal of replacing the old CIP 6/92 feed-in tariff system. For some years ahead, we will have a transition phase in which both CIP 6/92 and GC mechanisms will function side by side. Indeed, renewable plant owners had the option of applying to remain in the CIP 6/92 system and have the right to sell electricity for eight years to GRTN [Gestore della Rete di Trasmissione Nazionale], the Italian Transmission System Operator, and also to receive a feed-in tariff for the same period. GRTN is consequently then able to both sell that electricity to the electricity market and GCs to electricity producers, as those plants are eligible for Green Certification.

For those who had not exercised this option, which expired in 2000, there remained only the possibility of being qualified as a renewable plant to get Green Certification. Once the eight-year period is over, the plants under CIP 6/92 will no longer receive the feed-in tariff.

The Green Certificate Mechanism goes through a chain of four links: the obligations of electricity producers, Green Certification, a market, and verification.

Obligations

Starting from the year 2002, all electricity producers from conventional sources or importers must comply with an obligation to inject into the grid an amount of electricity produced with renewable sources equal to 2% of the total electricity produced or imported in the preceding year. In the period 2004–2006, this obligation will be increased by 0.35% each year.

The target can be met either by producing electricity directly from renewable sources or by buying GCs from other producers. By March 31 of each year, starting from 2003, each producer and importer has to deliver an amount of GCs equivalent to its obligation.

Qualification, Certification, and Registry

A renewable plant that entered into operation after April 1, 1999, is eligible for Green Certification. GRTN is responsible for the qualification of renewable plants. Technical features of plants that apply for qualification are examined by a commission; if the conditions are satisfied, the plant is qualified as a renewable plant and is eligible for Green Certification for eight years after the plant entered into operation.

GCs are issued by GRTN on the basis of electricity produced in the preceding year or the expected production of the current year and the following year. Each GC represents 100 MWh of electricity produced with renewable sources, and it is identified by the year in which the production took place. A GC related to one year can be used to comply with the obligation of the same year or the following two years.

GRTN also manages the GCs Registry, organized by accounts. GRTN opens an account for each renewable plant's owner and for each market participant under obligation; the GCs owned by a market participant are deposited in their own account.

Market

To comply with their obligations, producers and importers must deliver the equivalent amount of GCs to GRTN. Because of this obligation, transactions between participants with an amount of GCs higher than requested (long position) and participants with a need of GCs (short position) take place. The offers are represented by producers with long positions and GRTN for GCs related to CIP 6 plants. Producers or importers with short positions are on the demand side.

Even though bilateral contracts are allowed, GME [Gestore del Mercato Elettrico—the Italian electricity market operator] has a mandate to set up an organized market for GC trading. In March 2003, the GME market entered into operation. Since then, several sessions have been organized. They usually take place once a week in the first quarter, when the deadline to comply with obligations is approaching, and are monthly in the remaining part of the year, according to the market rules.

At the end of each market session, the list of the settled transactions is sent to GRTN for a registry update. Operators who have a bilateral contract must independently send a communication to GRTN with the transaction details. GRTN, after verification, updates the registry.

Verification

Every year, by March 31, all producers and importers must communicate to GRTN the number of GCs to be cancelled in order to comply with their obligations. GRTN cancels GCs from the operator's account in the registry. The Authority for Energy and Gas imposes a penalty for those who do not comply.

WHITE CERTIFICATE MECHANISM

Energy efficiency is another way to reduce GHG emissions. Italy has introduced two energy-efficiency decrees to provide an incentive for energy saving. According to these decrees, distributors of electricity and gas must achieve a saving target on the total consumption of their own customers. A yearly national target has been set for the period 2005–2009, measured in tonnes-of-petroleum equivalent (tpe).

Energy saving can be introduced through energy efficiency projects. Each project, after verification of the amount of electricity or gas or primary energy saved, is eligible for Energy Efficiency Certificates—EECs. Those EECs will be used by distributors to comply with their obligations, as they must deliver an amount equivalent to their target to the Authority for Energy and Gas. The introduction of a market mechanism for EECs allows other parties, such as energy service companies, to find participating in the market attractive and profitable. The Energy Efficiency Mechanism is very similar to the Green Certificate mechanism and is composed of obligations, a certification phase, a market, and a verification phase.

Obligations

Distributors of electricity and gas must achieve a saving target on the total consumption of their own customers, starting from 2005. The national target for the period 2005–2009 has been set either for electricity savings or for gas savings, and they are measured in tonnes-of-petroleum equivalent (tpe):

Table 1

Italian National Energy Efficiency Targets

Year	Electricity National target (Mtpe)	Gas National target (Mtpe)
2005	0.1	0.1
2006	0.2	0.2
2007	0.4	0.4
2008	0.8	0.7
2009	1.6	1.3

The national target is shared between distributors according to their relative weight. This weight is obtained by dividing the electricity or gas consumed by their customers with national consumption. To comply with their obligations, every year distributors must deliver to the Authority for Energy and Gas a number of EECs equivalent to their annual target.

Certification

A project that introduces energy efficiency is eligible for certification. The Authority for Energy and Gas has standardized several projects, indicating the amount of electricity, gas, or primary energy (measured in tpe) the project can save. Other nonstandardized projects can be evaluated on a case-by-case basis, and each will receive EECs for tpe saved.

Market

In the Energy Efficiency Mechanism, as in the GC mechanism, there will be obligations to be fulfilled and operators will have long and short positions in EECs. Even though bilateral contracts are allowed, GME has received the mandate to set up a regulated market with the same trading platform used for GCs, and with similar market rules.

Verification

By May 31 of the year following that of their obligations, distributors of electricity and gas must deliver an amount of EECs equivalent to their savings target. The Authority for Energy and Gas is responsible for verifying this compliance. For those who do not comply, a penalty will be applied.

Economics of Energy Efficiency Mechanism

Distributors that undertake an energy efficiency project can have three streams of income (see Figure 1).

The first comes from an agreement between distributors themselves and the end-user who will benefit from the project. As the total cost of the project is paid for by the distributor, it is usual to make an agreement where the end-user for a given period of time will pay back to the distributor part of the money saved as a result of the efficiency project. After that period, the end-users will receive the entire benefit of the project.

The second comes from a partial reimbursement of project costs paid for by the distributor and not covered from other sources. All distributors that undertake a project are eligible for tariff reimbursements for each tonne-of-petroleum equivalent saved with a project, up to the full realization of the target. The amount of reimbursement has been set at 100 euros for each tpe saved, and is obtained either by carrying out projects or delivering EECs bought from other distributors or ESCOs (Energy-Services Companies).

The third comes from the EECs sold in the market, when the distributor undertakes several projects and the total number of EECs obtained is more than that required to comply with their obligations.

Figure 1

EECs Market Mechanism

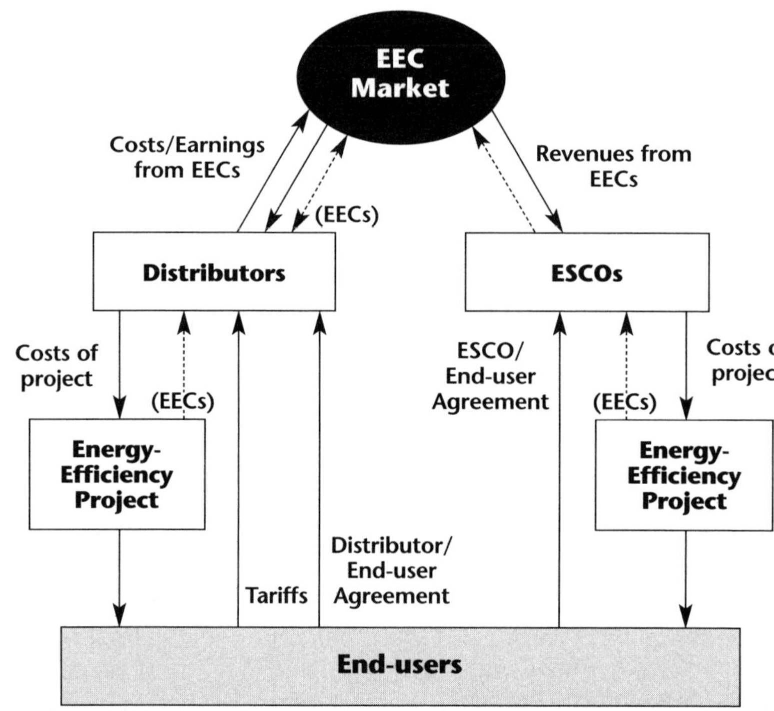

ESCOs can also make energy-efficiency projects without having an obligation to comply with. They find sources (usually with a "third-party financing" contract) to cover project costs and then obtain two income sources both from an agreement with end-users and from selling EECs into the market (see Figure 1).

As distributors can benefit from tariff reimbursements, the market mechanism allows this component to be transferred from distributors to ESCOs, when the project is undertaken by ESCOs. The price of EECs sold to distributors will take into account the value of the reimbursement the distributors will receive.

GME MARKETPLACE FOR GCS AND EECS

GME has organized a trading platform for GC trading. The same platform can be used for EECS trading when the market starts.

Operators who want to participate in the GC market will have to apply to become a market operator. After that, they can complete a registration form on GME's Web site. GME will validate the registration and will give a "company PIN" to the legal representative of the organization. With that password, it is possible to register one or more users through a user registration form. Each user will get a "user PIN" directly on the Web site at the end of the registration procedure.

In the GC market, operators can submit their bids and offers in a continuous trading environment, with automatic matching if price conditions are satisfied. A direct link with the GRTN Registry allows operators to sell only the GCs deposited in their own account, avoiding the risk of "double selling"; on other hand, buyers are requested to make a guarantee deposit on a GME account a day before each market session. GME sets the minimum amount to be deposited for each GC the operator is going to buy, and the operators may buy only the number of GCs equivalent to the total amount deposited.

At the end of each session, the transactions entirely covered by guarantee deposits will be settled and GME will pay the sellers. For those transactions not completely covered by guarantee deposits, buyers are requested to make an additional payment of the amount needed to fully pay the purchase price. Once this payment is made, GME will transfer money from its account to the sellers. If a buyer does not fulfil their payment obligations, the transaction is cancelled and a sanction will be applied.

As far as the EEC market is concerned, the admission procedure and the market features will probably be the same.

E.U. ETS AND BLACK CERTIFICATE TRADING IN ITALY

Denmark and the UK were the first European countries to introduce an Emission Trading System in 2000 and 2002, respectively. Now the European Directive 2003/87/CE has been approved introducing an E.U. Emission Trading System. According to this, since January 1, 2005, installations involved in activities listed in Annex I (see *Appendix 1*) must have a greenhouse gas emissions permit (the ability to measure and report emissions). Application can be made to the competent authority.

For each period (2005–2007, 2008–2012), member states develop a national plan to allocate the total quantity of allowances (Assigned Amount of Units—AAUs). The National Plan must be approved by the Commission of the European Communities. At least 95% of the allocation will be free of charge in the 2005–2007 period and at least 90% in the 2008–2012 period. Allowances can be traded within the European Community. By April 30 of each year, starting from 2006, the owner of each installation will surrender a number of allowances equal to its emission in the previous year. For those who do not comply with the obligation, a penalty is applicable (40 euros per tonne in the 2005–2007 period, 100 euros per tonne in the 2008–2012 period). The member states will organize a registry for allowances issued, traded, and cancelled. The Commission shall designate a central administrator to maintain an independent transaction log recording the issue, transfer, and cancellation of allowances. The Directive 2003/87/CE does not allow participants to comply with obligations delivering other credits obtained through JI [Joint Implementation] and CDM [Clean Development Mechanism] projects. A Linking Directive, that amends the 2003/87 in order to make credits coming from Emission Reduction Units (ERUs) and credits relative to Certified Emission Reductions projects (CERs) valid for complying with the E.U. ETS obligation, has been recently approved.

This Black Certificate Market will join the GC and EEC markets in Italy, even though it will be at a European level while the latter two markets will be national. For some participants, Green and Black markets (i.e., electricity producers) will overlap, and it can be useful to have the same market platform to trade their certificates. In this sense, GME is going to organize an emission rights market where both Italian and other European operators can buy or sell their black certificates, providing a complete offer of environmental markets.

Appendix 1

Activities Included in the E.U. ETS

Energy Activities

- **Combustion installations** with a rated thermal input exceeding 20 MW
- **Mineral oil refineries**
- **Coke ovens**

Production and Processing of Ferrous Metals

- **Metal ore** (including sulphide ore) **roasting or sintering** installations
- Installations for the **production of pig iron or steel** (primary or secondary fusion) including continuous casting, with a capacity exceeding 2.5 tonnes per hour

Mineral Industry

- Installations for the **production of cement clinker** in
  - ➢ rotary kilns with production capacity exceeding 500 tonnes per day, or
  - ➢ lime in rotary kilns with production capacity exceeding 50 tonnes per day, or
  - ➢ other furnaces with production capacity exceeding 50 tonnes per day
- Installations for the **manufacture of glass** including **glass fiber** with a melting capacity exceeding 20 tonnes per day
- Installations for the **manufacture of ceramic products** by firing (in particular, roofing tiles, bricks, refractory bricks, tiles, stoneware or porcelain) with
  - ➢ production capacity exceeding 75 tonnes per day, and/or
  - ➢ kiln capacity exceeding 4 m^3 and setting density per kiln exceeding 300 kg/m^3.

Other Activities

- Industrial plants for the production of
  - ➢ **pulp from timber or other fibrous materials**
  - ➢ **paper and board** with a production capacity exceeding 20 tonnes per day

Stefano Alaimo is *director* of the Environmental Markets Department at Gestore Mercato Elettrico (Rome), the Italian Electricity Market Operator. He can be reached at T: 39.06.8012.4656 and E: stefano.alaimo@mercatoelettrico.org.

CHAPTER 7

Expanding Markets for Renewable Energy Certificates

Progress and Challenges

By Edward A. Holt

More and more markets are recognizing that renewable energy certificates (RECs) are an easy way to differentiate environmentally preferred power sources.[1]

In states with renewable portfolio standards (RPS), retail electricity providers are required to include specified levels of renewable energy in their supply mix. The number of such RPS states continues to grow, and increasing regulatory compliance is established by ownership of RECs.

In addition to these compliance markets, REC use is growing in voluntary retail markets. Most consumers are not aware of the fact that RECs are often used to support marketing claims for green power.

RECs are also sold as a stand-alone (unbundled) retail product, unaccompanied by electricity delivery. Retail REC marketers now make it possible for all consumers to support renewable energy, including the roughly 60 percent that do not have access to a green power option from the electric grid. Unbundled RECs account for 27% (540,000 MWh) of renewable energy currently sold to commercial and institutional customers, according to the EPA's Green Power Partnership.

Finally, substantial amounts of both bundled and unbundled RECs are traded in the wholesale markets, which are largely invisible to consumers.

Although these markets for RECs are expanding, they also face barriers. Wholesale REC markets have few geographic constraints in theory, but retail markets are rather small and fragmented. For example, while RECs may trade freely in New England, some states within this region (Connecticut, Maine, Massachusetts, and Rhode Island) have different requirements for compliance with their specific mandatory RPS programs. In the west, regulations in Arizona, California, New Mexico, Nevada, and Texas require either that renewable generators be located in-state or that electricity from the power plants be delivered into the state. Such requirements limit the geographic scope and liquidity of REC markets.

Voluntary markets also vary according to how the retail electricity market is structured. Some states allow consumer choice of supplier and products, while other states maintain traditional utility franchise territories and rate regulation. In states with restructured markets, there may be one or more green power marketers, affording opportunity for REC sales; however, retail competition is not very strong. In regulated states, utility customers may have a green power option, but participation remains low. In both restructured and regulated states, markets are not broad enough or deep enough for REC marketers to rely on long-term voluntary demand.

This chapter is a synthesis of a panel discussion, "Renewable Energy Certificates: State of the Trading Market," at the 2004 GreenTrading Summit™: Emissions, Renewables & Negawatts. The panel, chaired by the author, included Karl Rábago, Energy Group Director at the Houston Advanced Research Center; Ed Mongan, DuPont Company's Director for Energy and Environment; and Steven Weisman, Director of the Green Power Program, Massachusetts Technology Collaborative.

REC CERTIFICATION AND VERIFICATION

REC standards and definitions are important to clarify REC ownership and claims. As with bundled green power, REC marketers need to substantiate that renewable generators actually produce the RECs claimed, that they actually own the RECs sold, and that they make accurate marketing claims. The ability to prove these facts adds to the credibility of the REC market.

In addressing these questions, Karl Rábago described the Green-e Product Standards for Tradable Renewable Certificates (TRCs). The Green-e TRC Standard was created through a national stakeholder process, and recommendations were approved by the Green Power Board—an independent policy board chaired by Rábago. The Green-e Standards are implemented by the staff of the nonprofit Center for Resource Solutions (CRS).

For REC product certification, the Green-e TRC Standard includes the following:[2]

- Only RECs from new renewable energy facilities will be certified.
- Eligibility is determined using a national definition of renewable energy.

- Certified RECs must include all environmental attributes associated with the generation.
- Certification is directed at voluntary REC markets; Green-e does not certify RECs used in mandated or RPS markets.
- Geographic disclosure (the location of the generator) is required because RECs from a given generator can be sold virtually anywhere.
- No double counting is allowed, meaning that a REC used for one purpose may not also be used for another purpose. For example, a REC used to satisfy an obligation such as an RPS may not also be used to support a marketing claim in a voluntary market. This implies that a REC cannot be sold to more than one party for final use.
- Relevant state regulators must be notified of a REC transaction when a utility is involved, so they can consider the revenue implications.
- Environmental disclosure and contract information must be provided to consumers.
- Green-e will perform an annual compliance review and verification audit.

Green-e verifies that these standards are met. In addition, Green-e conducts marketing compliance reviews to consider what REC claims are being made, whether the claims are legal and easily understood, and whether consumers were properly informed about the product. In addition to certifying that these standards are met, Green-e also verifies that the same number of RECs was supplied as sold, and that the types of RECs supplied were consistent with the claims made to consumers.

According to Rábago, in 2002 Green-e certified a total of 1,926,000 MWh of renewable energy supply, representing 58 percent of the total U.S. retail market. Of these voluntary renewable energy sales, 1,480,000 MWh were sold (bundled) in competitive electricity markets, 79,200 MWh were sold in utility green pricing markets, and 367,000 MWh were sold as unbundled RECs. Sixty renewable energy products were certified (including 27 REC products) involving the participation of 98 marketers (including 21 REC marketers).

The range of marketers selling Green-e certified products is illustrated in Figure 1. Active retail marketers and wholesale marketers sell unbundled RECs while brokers trade RECs, and numerous utilities and marketers sell REC-based products. Most renewable energy transactions today now involve RECs at the wholesale or retail level.

Figure 1

Suppliers of Green-e Certified Products[3]

REC TRACKING AND ACCOUNTING

Electronic REC tracking and accounting systems are being developed to assist in verification and certification. These systems are being designed and implemented to perform the essential functions of issuing certificates, tracking changes of ownership, and retiring certificates when they are used to support compliance or voluntary claims. Tracking systems thus create a record of generation that contains a variety of information useful or necessary to determine the REC's eligibility in compliance markets or its desirability in voluntary markets. Therefore, the tracking systems serve as essential verification tools to help prevent double-selling of RECs and support credible REC markets. Figure 2 shows certificate tracking systems in operation and those under development or discussion.

REC Tracking Systems that are in operation today are found in New England (NEPOOL Generation Information System) and the states of Texas, Wisconsin, and the Canadian province of Manitoba. Stakeholders are currently developing design criteria for the PJM Generation Attributes Tracking System (GATS) and the Western Renewable Energy Generation Information System (WREGIS), a regional effort including 11 Western U.S. states, 2 Canadian provinces, and Northern Baja. Also, tracking systems are under discussion in

Ontario, New York (as part of the RPS rulemaking), and in the Upper Midwest states of Minnesota, North Dakota, South Dakota, Iowa, and Wisconsin.

Rábago asserted that the equivalent of a financial banking system is needed for REC tracking systems. State, provincial, regional, or national REC "issuing bodies" need common practices for handling accounts and for

Figure 2

Certificate Tracking Systems in the North America

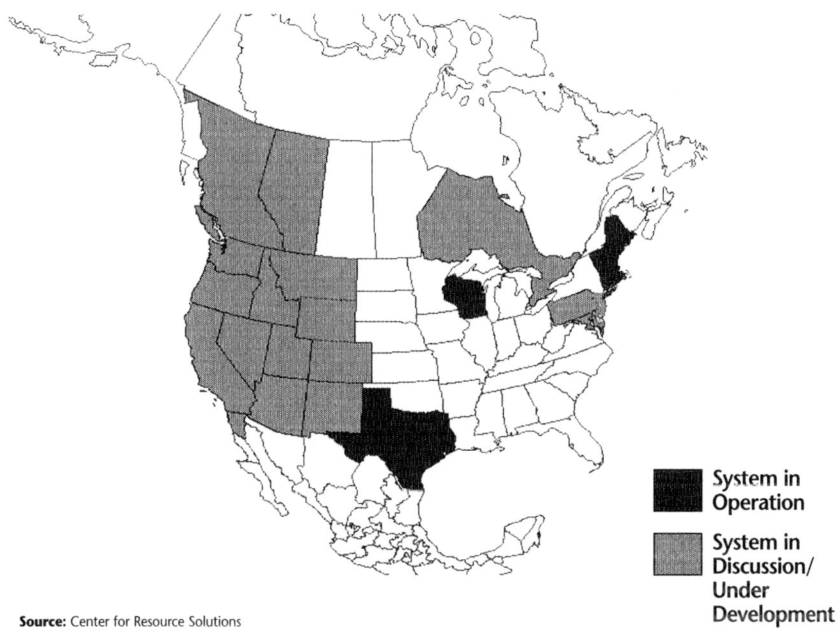

Source: Center for Resource Solutions

transferring RECs, particularly for imports to and exports from one system to another. A North American Association of Issuing Bodies (NAAIB) has been established to promote compatibility among systems and to provide credibility to the emerging REC markets.

Issues of concern in the design of tracking system design include:[4]

- certificates for all generation or renewable generation only
- geographic scope
- institutional support for design and operation

- system coordination and imports / exports
- disaggregation of attributes
- data acquisition
- market-making responsibility
- cost allocation and fees
- development process issues

With these tools in place—verification and certification supported by regional REC tracking systems—Rábago sees the use of RECs expanding dramatically. A growing number of corporations and large institutions are buying RECs, including the Commonwealth of Pennsylvania (10 percent of total electricity use), the State of Oregon, the U.S. Army, U.S. EPA (Environmental Protection Agency), more than 30 colleges and universities in Pennsylvania, Nike World Headquarters, White Wave (makers of Silk soy milk), Interface Fabrics, and DuPont Company. While their motivations may be as varied as the companies themselves, the following example serves as a case study of the multiple dimensions of the choice for renewable energy in the form of RECs.

CASE STUDY: DUPONT COMPANY

DuPont represents a large corporation's effort to pursue green power and the purchase of renewable energy resources on a voluntary basis. Founded in 1802, this 200-year-old company has annual revenues of approximately $25 billion and a total of 59,000 employees at 180 facilities in 70 countries. Key markets include aerospace, automotive, construction, crop protection, electronics, nutrition and health care, safety, and security.

According to Ed Mongan, one of the key drivers behind DuPont's interest in renewable energy is the corporate commitment to sustainable growth, defined as creating shareholder and societal value while decreasing its environmental footprint along the value chain. Among other things, this vision of sustainable growth means reducing nonrenewable sources of energy and raw materials as well as waste and emissions.

In 1999, DuPont set some aggressive goals, committing the company to:

- Reduce greenhouse gas (GHG) emissions by 65 percent between 1990 and 2010;
- Hold energy usage flat, on an absolute basis, from 1990 to 2010; and
- Supply 10 percent of total energy needs from cost-competitive renewable energy sources by 2010.

Alternative fuels—including enhanced hydropower, biogas, landfill gas (used directly as fuel or converted to electricity), paper and wood chips to steam, and wind and solar—are a critical part of reducing the company's GHG emissions.

Reducing energy costs is a critical incentive alongside DuPont's commitment to reduce its environmental footprint. By early 2004, the company has identified $5 million worth of annual savings through its biogas projects alone. Seeing competitive opportunities in wind, biomass, and landfill gas, DuPont is motivated to spur innovation and development of these renewable sources.

While DuPont's main renewable energy sources are biomass projects in the Midwest, the company has made a three-year commitment to purchase RECs to achieve some of its renewable energy goals. Working with the World Resources Institute's (WRI) Green Power Market Development Group,[5] DuPont has purchased 170,000 MWh worth of RECs in 2003, the equivalent of nearly 20 average MW of generating capacity and about 1 percent of its total global energy consumption.

The separation of environmental attributes from commodity electricity makes RECs a more cost-effective and flexible method of acquiring renewable energy than traditional bundled green energy products. In purchasing RECs, DuPont looks for reliable providers, fixed and competitive prices, significant environmental attributes, third-party verification, and—most important—newly installed renewable capacity, since part of the company's objective is to stimulate renewable energy growth and development. Environmental benefits notwithstanding, Mongan believes that renewables must be cost-competitive with, if not cheaper than, fossil fuel–based alternatives if they are to solidify and expand its role in our energy future.

Based on its experience with RECs, DuPont supports the following recommendations for the future development of RECs markets:

- national registration and tracking system for RECs
- clear rules regarding ownership of environmental attributes
- standard protocol for calculating avoided emissions
- ensuring a role for RECs in emissions trading markets, and
- incorporating the long-term fuel-cost benefits

LONG-TERM REVENUE SUPPORT FOR RECS

Even with the many societal and environmental benefits, renewable energy development requires continuing financial support. Without long-term contracts from credit-worthy buyers, renewable energy developers have difficulty in securing financing for their projects. Where monopoly markets and regulated rates of return still exist, vertically integrated utilities are the logical long-term buyers if they are interested. In more competitive markets with attendant higher risks, it is much harder to find a player to take on this role. Third parties have emerged to help developers secure project financing, including the innovative Massachusetts Green Power Partnership (MGPP) created by the Massachusetts Technology Collaborative.

The Massachusetts Technology Collaborative is responsible for overseeing the Renewable Energy Trust, a dedicated fund created by Massachusetts's 1998 restructuring legislation. The Trust has collected approximately $25 million annually for renewable initiatives and projects. Its objectives are to increase the supply of and demand for energy from clean sources, promote the development of a vibrant Massachusetts renewable energy industry, and maximize the benefits to Massachusetts rate payers.

Steven Weisman, director of the Green Power Program, discussed the structural obstacles for renewable energy development in New England, where most of the electrical load is from restructured markets in which utilities no longer own generation and are therefore no longer in a position to enter into long-term contracts. The responsibility for resource development has been shifted to third-party suppliers wary of entering into long-term contracts for fear that their costs will not be recovered. Even with a mandatory RPS, few entities are willing to make intermediate-term five-year commitments for energy or RECs, and most companies generally satisfy their obligations with year-to-year purchases. These short-term purchases, however, do not facilitate the development of new projects.

Since most banks or other financial institutions are unwilling to take revenue risk on merchant plants, there are very few creditworthy entities in the REC market today. Further limiting is the factor of project size; most New England projects are small (with wind projects of less than 40 MW), making them less appealing to financial markets looking for larger investments.

The Renewable Energy Trust created the MGPP as one way to provide REC revenue certainty to developers through long-term contracts whereby the Trust helps manage some of the market risks in future REC value and demand.

The MGPP has solicited proposals for the funding of new projects from renewable generators or project developers, as well as from power marketers and other purchasers of energy and RECs (with a renewable generator as co-proposer). Projects are evaluated against economic and technical criteria to minimize the Trust's exposure to risk and maximize value to Massachusetts ratepayers. These criteria include project strength, likelihood of success, and the overall value created.

Under the MGPP structure, the Trust contracts with developers either for (a) REC purchase agreements or (b) put/collar option contracts. These contracts can be up to 10 years in length, for a period that does not need to include the first 10 years of a project's life. The contract obligation can extend through 2021 but not beyond. Project awards were restricted to any new construction or incremental generation eligible for the Massachusetts RPS (with new hydro also acceptable). Moreover, projects must be in commercial operation by December 31, 2005, and must sell RECs to Massachusetts customers. The Trust contracts cover REC revenue only and assume that energy revenues will be recovered through the electricity markets.

In the first award group (and there has been only one as of March, 2004), the MGPP asked proposers to bid the product, the price, and the term. The Trust and the successful bidders enter into contract, and then, the Trust escrows funds to cover its obligation. As a facility is financed and built, the Trust purchases RECs according to the agreement. The Trust's risk is therefore tied to the actual production from which the Trust can resell the RECs for RPS compliance or for the voluntary Green Choice program. Alternatively, the Trust might sell REC contracts to a third party. If the project is not built or if the option is not exercised, the escrowed funds will be released.

In the first award group, the MGPP received eleven proposals and funded five projects. The awardees represented a broad range of technologies including biomass, hydro, landfill gas, and wind generation, and there was a combination of REC purchases, put options, and collars. In this round of awards, the marketplace diversity intended by the Trust's mandate was promoted. In total, $32.2 million were awarded (in nominal dollars) but only $20.2 million are escrowed in anticipation of the growth and future value of these funds. Nearly 99 MW of renewable capacity will be installed, with expected production of 486,000 MWh (in RECs) for the New England/Massachusetts market and with 185,000 MWh specifically designated for the Massachusetts RPS. Figure 3 provides a description of the awarded projects:

Figure 3

Massachusetts Green Power Partnership Round 1 Awardees

Project and Applicant	Tech.	Location	Form	MW	Present Value Exposure ($)
Berkshire Wind, LLC (sub of DISGEN, Inc.)	Wind	Hancock, MA	Purchase	13.5	5,176,765
Hoosac Wind, LLC (sub of enXco, Inc.)	Wind	Florida, MA Monroe, MA	Option	30.0	10,042,331
Greater New Bedford LFG (Commonwealth Resource Management Corp.)	Landfill Gas	Dartmouth, MA	Option	3.3	1,990,795
Pepperell Hydro, LLC (sub of Swift River Co.)	Hydro	Pepperell, MA	Purchase	1.9	607,959
Northern Wood Power—Schiller (Public Service Company of NH)	Biomass	Portsmouth, NH	Option	50.0	2,378,930
			Totals	**98.7**	**20,196,780**

Four of the five contracts were signed as of March 12, 2004, and the funds all went into escrow within 48 hours of closing. The progress of the awardees will be monitored toward the December 31, 2005, construction deadline.

The first round of the MGPP has demonstrated that developers are interested in long-term contracts provided that price security is to some extent guaranteed. The Trust has also learned that later-year (i.e., years 5 to 15) price support is of greater concern for most bidders than near-term sales because they are more confident of finding buyers in the near term. In addition, put options as guarantees seem more attractive to developers than commitments to REC purchases, since they can have the guarantee of the floor if the market turns out to be greater than was expected.

Looking ahead, the Trust has begun the process of offloading REC contracts in order to release committed funds. The second round of the MGPP is scheduled to take place in the fall of 2004. However, realistically, the Renewable Energy Trust does not have enough money to provide revenue security to large projects, much less to the total of Massachusetts renewable energy demand. Therefore, it is imperative for large creditworthy companies such as the load-serving entities or large end-users to step in and make long-term commitments to finance projects.

EMERGING ISSUES IN REC MARKETS

Clearly, the REC markets are making progress. More states are creating demand through policy mandates and through growing green power options. Important tools are being created through REC product certification to a voluntary standard and through the development of regional and state REC tracking systems. Large corporate and institutional REC buyers are entering the market, adding credibility and significant demand. Renewable energy funds are beginning to facilitate long-term revenue streams to new renewable developers, through purchases and guarantees that remove reduce revenue risk. In this developing picture, some important issues warrant monitoring and require clarification.

First, what is the role of regulators? Since renewable energy certificates (RECs) are sold separately from electricity, one might think they are not regulated. Yet, RECs will certainly come under regulatory scrutiny when they are used by regulated companies or for compliance with mandated programs. For example, REC sales affect revenue for utilities regulated for rate of return and, therefore, electricity rates. If RECs are required for RPS compliance, or to substantiate marketing claims, regulators will be setting the rules. Also, RECs may be used to help verify consumer product labels for electricity that are required in nearly half of the U.S. states. In addition, there may be disputes about REC ownership where contracts are not explicit about RECs. In such cases, companies have begun to look to state utility commissions to clarify REC ownership. This is particularly true for Qualifying Facility contracts under the 1978 Public Utility Regulatory Policy Act.[6]

A second REC issue is the tension between a desire for larger, more liquid REC markets, on the one hand, and the desire for local benefits on the other. Many renewable energy generators and REC traders would like to see large regional, if not national, REC markets. Generators in particular would like to be able to reach more potential buyers, and traders want more liquidity to make trading easier. More buyers and sellers also yield more competition and lower REC prices, benefiting consumers. Contrast that with the possible motivations of a state adopting an RPS. Some states may wish to restrict RPS eligibility to renewable generation located within the state in order to encourage local development of renewables, protect local resources from competition, and preserve local economic benefits.

The tension between these two desires needs to be managed. For one thing, limiting markets for imported (out-of-state) RECs may violate the U.S. Commerce Clause and encourage traders to bypass tracking systems, which could create opportunities for double-selling. Currently, state regulators are trying to find a balance between capturing local benefits and supporting more competitive and open markets. Some states are considering a reciprocity approach in which one state would accept certificates from another if the latter does the same. Other states require that imported RECs be accompanied by an energy delivery, and since power distribution does not generally conform to political boundaries, this requirement can really be meaningful only in the context of a power pool perspective rather than within the geographic area of a state.

A third issue is that renewable energy has only very limited access to the environmental markets. Using renewable energy to achieve emissions reduction goals is challenging and raises additional issues. More fundamentally, the question that needs to be addressed is whether or not renewables are at all eligible to earn emission allowances or credits that can be traded for compensation. This is important to renewables developers and owners because they are interested in the potential revenue, especially if certain federal tax incentives for renewable energy development remain expired or are renewed for only a few additional years. The uncertainty about renewables' participation in emissions markets, as well as the operations of the emissions markets themselves, must be resolved for this market to benefit renewable energy development. At this interface between the REC and emissions markets, both utility regulators and air regulators need more awareness and understanding of the implications of the rules they promulgate. For now, emissions markets tend to be small state-by-state markets, and operate without conferring long-term value that can be taken to a bank for project financing.

A final issue relates to disaggregation of REC attributes. Each REC is an aggregation of attributes, such as the fuel type, generator location, vintage of generation, and emissions of individual pollutants. Some stakeholders are beginning to advocate for disaggregating or stripping off individual emissions attributes and converting them (if eligible) into emission reduction allowances that may be sold in emission cap-and-trade programs. This disaggregation has the potential to create multiple revenue streams to the owner. For example, if a wind generator could obtain a NOx emission allowance under a state set-aside program, find a buyer for its CO2 emission-free attribute, and use what remains of the REC to satisfy compliance with a state RPS, it might come out ahead financially.

Most of the market development so far, however, is based on the inclusion of all attributes—a sense that the REC should be "whole." Stakeholders with this perspective worry about the credibility of retail energy markets if RECs are sold without some of their essential constituent attributes. It would be hard to explain to consumers what they are getting if the emissions benefits have been split off and sold separately to another party.

Many stakeholders view disaggregation and selling individual attributes to different parties as double-selling, though advocates of disaggregation would argue that by defining a REC as proof that a renewable MWh has been generated, the emissions attributes are free to seek their own markets.

Ultimately, whether to allow disaggregation of attributes is a state-level policy question. Although many state energy regulators are sympathetic to concerns about retail markets (and hence, might discourage disaggregation), it is state air regulators that are being encouraged to establish mechanisms to award emission allowances to renewables. Air regulators are not responsible for the credibility of retail energy markets, and the two sets of regulators are not addressing the issue in a coordinated way. The issue is new enough not to have been addressed in rulemaking. Therefore, regulations are silent on this issue except in a few cases (California, Connecticut, New Jersey, New York, and Texas) that call for all environmental attributes to be included with the REC.[7]

The resolution of this issue may revolve around whether policy makers expect additional benefits from each policy they adopt. If policy makers expect additional or different benefits from an RPS, compared to an air-quality program capping emissions, then disaggregation might not be allowed. However, if policy makers adopt an RPS to achieve the same goals as the air-quality program, then there may be no reason to prevent using the same MWh to achieve both purposes. Currently, though, a policy vacuum exists, making the rules unclear.

As noted at the beginning of this chapter, a growing number of REC markets are fragmented and illiquid. If RECs stakeholders do not reach consensus about REC definitions and instead pursue widely divergent market rules, we risk further market fragmentation and increased confusion. To avoid this, we need greater policy clarity and education for policy makers on critical intertwining issues. Energy regulators should be more aware of energy policy implications for air-quality program rulemaking, and air regulators should be more aware of environmental policy implications for energy program rulemaking.

---

[1] A useful introduction to RECs may be found in Peter Fusaro and Marion Yuen, *GreenTrading™: Commercial Opportunities for the Environment* (New York: GreenTrading, Inc., 2004), chapter 6, "Renewable Energy Certificates: The State of the Market."

[2] To read the actual TRC Standard, go to http://www.green-e.org/pdf/trc_standard.pdf.

[3] Figure is excerpted from Karl Rábago's presentation at the Third Annual GreenTrading Summit™: Emissions, Renewables & Negawatts (March 23, 2004).

[4] For more details about tracking system design, see National Wind Coordinating Committee, "Design Guide for Renewable Energy Certificate Tracking Systems" (2004) at http://www.nationalwind.org.

[5] The Green Power Market Development Group, a collaboration of 12 leading corporations and the World Resources Institute, is dedicated to building corporate markets for green power. See http://www.thegreenpowergroup.org.

[6] More details about these and other regulatory issues can be found in the *Regulator's Handbook on Tradable Renewable Certificates,* at http://www.resource-solutions.org/RegulatorHandbook.htm

[7] California PUC, Order Instituting Rulemaking to Implement the California Renewables Portfolio Standard Program. Rulemaking 04-04-026. Opinion Adopting Standard Contract Terms and Conditions, mailed May 17, 2004. 'Environment Attributes' means any and all credits, benefits, emissions reductions, offsets, and allowances, howsoever entitled, directly attributable to the generation from the Unit(s). Connecticut DPUC, Promulgation of Regulations for licensing of electric suppliers and administration of renewable energy portfolio requirements: "Any electric supplier that seeks to demonstrate renewable energy portfolio standard compliance by participating in a renewable energy trading program shall have exclusive ownership of all renewable energy and environmental attributes from such trading program that are associated with its renewable energy sources." New Jersey Administrative Code 14: 4-8: "Renewable Energy Certificate or REC means a certificate representing the environmental benefits or attributes of one megawatt-hour…" New York, in its solicitation of renewable energy attributes for compliance with its RPS, uses a definition of attributes that is similar to California's. Texas ERCOT Protocols, Section 14: Renewable Energy Credit Trading Program: "A Renewable Energy Credit is a tradable instrument that represents all of the renewable attributes associated with one MWh of production…"

Edward A. Holt is *president* of Ed Holt & Associates, Inc., an independent consultancy in renewable energy, green power marketing, and certificate trading. He can be contacted at **T**: 207-798-4588 and **E**: edholt@igc.org. The author would like to thank Hoi Ying So, *research analyst* at Global Change Associates, for her help in preparing this chapter.

CHAPTER 8

Wind Energy: Promoting a Cleaner Energy Future

By Mark M. Little

Today, energy portfolios increasingly are shaped by local environmental and economic requirements. On a global basis, fuel mix decisions are influenced by a common set of factors: security of supply, commodity price volatility, infrastructure investment, technology risk, and environmental impacts.

In the foreseeable future, the foundation of most energy portfolios—fossil fuels—will continue to power the world's economies. Within these coming decades, however, portfolios will increase their shift toward a greater mix of renewable solutions. While fossil fuels offer very favorable energy densities and an existing distribution/generation infrastructure, their finite nature and price volatility in combination with increasing environmental concerns and source-country political dynamics have fostered the development of alternative energy solutions. In fact, some studies project that by the middle of this century, 30 to 50 percent of the world's future electricity will be generated by means of a mix of renewables, including wind power. A shift is now underway.

Although the rate at which the shift will occur is debatable, recent events support a sustained migration to renewables. The ratification of the Kyoto Protocol, for example, set the stage for monetization and trading of emissions credits. This, in turn, has set the stage for additional bankable cash streams on which to justify the financial viability of renewable energy development. At the same time, the increased demand from emerging economies, such as China and India, on energy and infrastructure commodities (including steel, oil, and gas) continues to push fossil fuel prices upward with no significant reductions expected in the near term.

Concurrently, improvements in cost, reliability, and grid integration have addressed many of the historical concerns associated with a greater reliance on renewable energy resources. Today, wind energy has emerged as one of the more viable utility-scale renewable energy solutions. The benefits of wind energy are obvious: It does not generate air or water emissions, it does not produce hazardous waste, it does not deplete natural resources such as coal, oil, gas, or water, nor does it cause environmental damage through resource extraction and transportation.

While wind energy is not the entire answer to the world's air pollution problems, it certainly can offset a significant amount of the emissions currently produced by traditional power generation methods. It also is becoming increasingly cost competitive; already it is competing with fossil fuel generation in some markets.

While wind energy has many benefits, there are also challenges that must be considered. Since the wind doesn't blow all of the time, this intermittency as well as line losses that occur when wind plants are located at considerable distances from electricity load centers, can impact negatively wind power's economics. A comprehensive policy package that clearly addresses transmission grid extension and the monetization of environmental benefits is necessary, as is a firmly established goal to explicitly promote the development of renewable resources.

At the end of the day, energy strategy plays out in the portfolio of fuel choices and must reflect a more thoughtful balance of the environmental, economic, security, and infrastructure costs today and into the future.

WIND POWER

The concept of harnessing the power of the wind has been an idea long explored, but with mixed results. It was not until the energy crisis of the mid-1970s that sustained efforts from the public and private sectors gradually brought wind-power technology to a level where it is considered commercially viable.

Wind energy's environmental benefits are substantial. This year, for example, the American Wind Energy Association (AWEA) estimates that U.S. wind energy plants will generate 16 billion kilowatt-hours (kWhs). This will be accomplished without generating air or water emissions and without causing environmental damage through fuel extraction. By reducing the need to burn traditional fossil fuel sources, wind power plants in the United States alone will offset 10.6 million tons of carbon dioxide, 56,000 tons of sulfur dioxide, and 33,000 tons of nitrogen oxides that otherwise would be released into the atmosphere this year.

THE TECHNOLOGY

Many innovations have been applied to wind technology over the past few decades. After considerable public and private investment and research and development (R&D), a consensus has been reached on the basic architecture of a modern wind turbine: an upwind, horizontal axis, variable speed, three-bladed rotor mounted atop a tubular tower.

Along with increases in scale, height, and rotor diameter, a number of additional innovations associated with blade composite materials, blade aerodynamics, systems integration and controls, power electronics, and grid robustness have enabled wind turbines to capture more efficiently the kinetic energy of the wind.

These improvements in technology have decreased wind power's cost of energy to a level where it is now economically viable and is, therefore, in a better position to compete within the traditional fossil fuel mix. Figure 1 provides a high-level view of the turbine size (kW rating) versus cost-of-energy ($/kWh) over past decades.

Figure 1

Progressive Development of Wind Energy

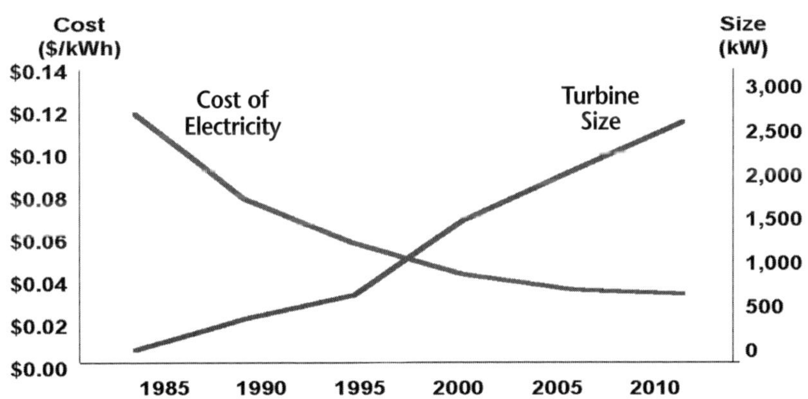

Source: GE Energy Analysis

HISTORIC ROADBLOCKS: GRID INTERACTION AND AVAILABILITY ISSUES

The variability of power production from wind turbines has been one of the major challenges facing grid operators in today's deregulated power markets. With a standard procedure that calls for scheduled energy deliveries every 10 minutes, wind energy has been subject to financial penalties by many grid operators because it cannot conform to these standards based on the attributes of thermal power units. To address this concern, last year the California Independent System Operator (Cal-ISO) launched a new scheduling system for wind energy that is viewed as a model for grid managers by the Federal Energy Regulatory Commission (FERC). The new system currently is being deployed in a pilot project under which wind generators net the difference between forecasted and delivered electricity every hour instead of every 10 minutes. A monthly financial settlement process then nets deviations from all energy production forecasts across hourly intervals—all at the weighted average electricity price for the month.

The wind energy industry prefers to rely upon a monthly time frame for settlement, rather than every 10 minutes. A wind project's electricity generation follows fairly consistent seasonal patterns. Hourly and daily deviations in the variability of wind largely fade into the background when power production is netted over the course of an entire month instead of every 10 minutes. Those participating in this Cal-ISO program are exempt from previous penalties relating to meeting their scheduled energy deliveries. In return, each participant is required to provide real-time meteorological and energy production data. This data then will be used in future forecasts of energy production for each individual wind project. This database will allow the wind industry, as well as the Cal-ISO, to understand better each wind project's power production profile.

Historically, wind turbines also have posed challenges for grid operators. Wind generators were often the first type of generator to trip offline during a grid fault event, leading to the loss of large blocks of power from wind projects. Now, with the emergence of "wind farm management controls," GE's wind turbines can moderate the voltage and Volt-Amp-Reactive (VAR) output from the wind power plant and provide stability to the grid.

With recent technological innovations, GE's wind turbines now can shift into a "low-voltage ride through" mode of operation where they can remain on line and generate electricity during a fault, riding through voltage declines as much as 30 percent for 600 milliseconds.

Some of the benefits of variable speed technology and VAR innovations are:

- Reduced power and torque excursions;
- The ability of the rotor to operate at maximum aerodynamic efficiency, providing increased power output;
- Lower loads on key components, which reduces part replacement costs;
- The ability to provide remote control of voltage and power factor for utility grid integration; and
- The stabilization of weak grids, making wind power more beneficial to utilities.

Despite the innovation, the variability of wind resources still limits the amount of capacity a wind turbine can deliver to the grid. Today, GE wind turbines that are located in the best terrestrial sites are operating at a 40-percent capacity factor, a dramatic improvement from first generation technology. A better measure of the utility of modern wind turbines, however, is reliability and availability data. GE's fleet of state-of-the-art wind turbines are posting availability figures in excess of 96 percent and are frequently achieving more than 98 percent availability in challenging environments. In other words, when the wind resource is available, these machines generate electricity 96 to 98 percent of the time.

WIND MARKET EVOLUTION

As we look at the commercial side of wind energy, it is apparent that stable energy policy support (well exemplified by northern Europe) has been very successful in nurturing the technology and operational advances over the past decades. Through 2002, Germany and Denmark accounted for a good portion of installed wind turbine generating capacity. Because of the sporadic and unpredictable nature of the U.S. Production Tax Credit, most U.S. wind turbine OEMs (original equipment manufacturers) and suppliers have found it difficult to justify sustained investment in technology or capital equipment. As a result, most of the wind energy supply chain has established roots in northern Europe.

With the emergence of more economic wind power technology, the passage of Kyoto's emissions reduction targets for carbon dioxide (CO_2), and the upward price pressures on fossil fuel, more and more countries are adopting renewable portfolio standards (RPS). The RPS proactively requires a set amount of renewable energy to be produced by a given date. The European Union adopted "20 percent by 2020" as a policy framework to meet Kyoto emissions targets. Most of the world, including 18 states in the United States, followed their lead by enacting an RPS. In fact, the number of countries actively building a wind energy infrastructure has evolved quickly over the past several years due, in large part, to some form of an RPS.

To illustrate the evolution, we've taken a snapshot at the global market at two intervals: 2000 and 2008. In 2000, there were approximately 15 countries actively developing wind power projects with combined sales of approximately \$5 billion (Figure 2). Only a handful of countries had significant installations (those that are highlighted generated greater than 100 MW), with the dominant countries being Germany, Spain, the United States, and Denmark.

Figure 2

The Global Market for Wind Energy (2000)

2000: ~15 Countries

In 2004, there are 45 countries actively pursuing wind with the major markets. The market size is approximately $8 billion, with Germany, Spain, and the United States continuing to be the largest segments.

As shown in Figure 3, the 2008 forecast highlights approximately 60 active countries, with most of the industrialized world building wind energy into their portfolios. The market is expected to be around $12 billion with the offshore segment accounting for around 20 percent.

Figure 3
The Global Market for Wind Energy (2008)

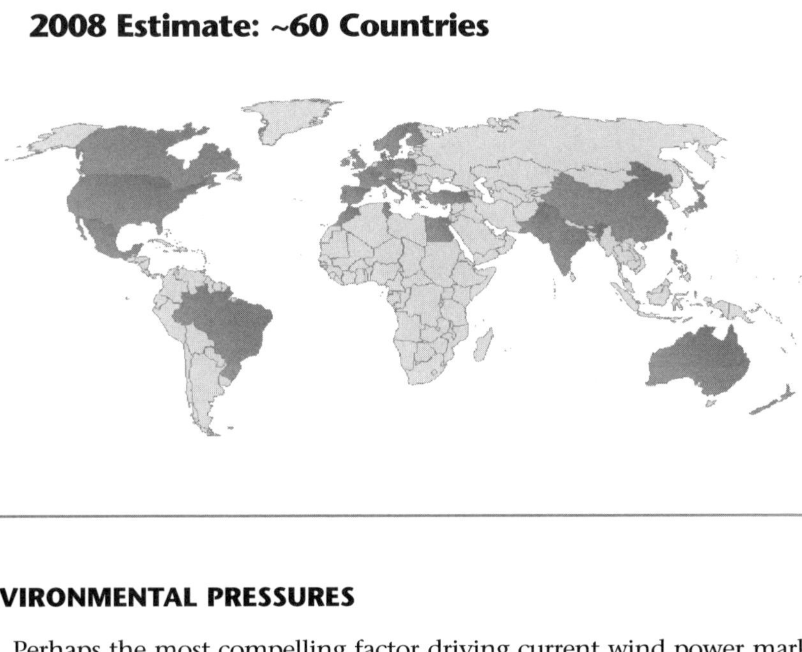

ENVIRONMENTAL PRESSURES

Perhaps the most compelling factor driving current wind power markets at the portfolio level is the growing environmental pressure. With the recent adoption of the Kyoto Protocol by Russia, the stage is set for global monetization and management of CO_2 and other emissions. CO_2 comprises 80 percent of the world's total current greenhouse gas (GHG) inventory. Roughly half of these CO_2 emissions come from electric power generation. Therefore, it is expected that the focus on balancing the energy portfolio fuel mix between fossil fuels and renewables will increase dramatically in the near future.

ENERGY DEMAND ON THE RISE

The world energy picture is set for some fundamental changes in the coming decades. According to the United Nations and the Population Reference Bureau, the world population will hit about 8.7 billion in 2050 and over 10 billion by 2100. Increased standards of living, emerging economies, and increases in the development of mega-cities will continue to increase energy demand significantly.

ENERGY SECURITY

Today, both Europe and North America are large importers of energy. Europe imports approximately half of its energy supply from Russia and the Middle East, whereas the United States imports about one-third of its energy. This dependence on foreign sources for energy can create economic instability if supply disruptions occur. With one of the largest wind energy resources in the world, the United States can extend the life of its oil and natural gas reserves by tapping into this renewable fuel. The modular nature of wind also offers national security benefits in terms of distributing power generation facilities throughout the country and negating the need to import fuels. For those geographic areas featuring strong wind and solar as well as other renewable resources, renewable energy can be a viable solution to reduce the demand for imported energy.

FOSSIL FUEL PRICE VOLATILITY

Since it is relatively clean and readily available, there has been a dramatic increase in the demand for natural gas over the past decade. In 2004, prices for natural gas increased by approximately $1.50 per million British thermal units (BTUs). Economists now predict natural gas prices will stay in the range of $5.50 per million BTUs for the foreseeable future. Yet, efforts to import liquefied natural gas (LNG) into the United States have been hampered by security and environmental concerns in local communities.

Prices for coal, which is the other primary fuel used for power generation in the United States, are also increasingly volatile. Because it is estimated that the United States has a domestic supply of coal that would last 200 to 250 years, significant investments are being made to make cleaner coal a reality during this decade. Yet pollution controls and other technological add-ons increase the cost of one of the world's least expensive electricity fuels.

INFRASTRUCTURE DECISION POINTS

China, India, and Mexico are among the emerging economies that are at a critical juncture in terms of planning for their energy futures. Just as cell phones are displacing traditional telephonic infrastructure, we would expect to see emerging economies make more creative energy decisions, including a longer-term view of renewables and an increased use of system-level hybrid solutions (such as wind/hydro and nuclear/hydrogen).

By its very nature, wind power can reduce risks in any given utility energy portfolio if one considers the natural hedge against volatile fossil fuel prices. An investment portfolio that contains fluctuating stocks as well as steady and secure bonds is exposed to less volatility than a simple stock portfolio. The same is true in power generation: Relying only upon fossil fuels that have witnessed such incredible volatility in recent years is not a prudent power supply strategy.

Today, wind power is integrated successfully into larger energy portfolios around the world. Clear evidence of this wind-power success story can be seen in some areas of northern Europe where wind energy supplies between 10 and 25 percent of the electricity needs of entire regions.

ROLE OF GREEN TRADING CREDITS

Regulatory structures and compliance regimes vary widely from state to state and country to country. Yet there is a consensus about the efficacy of the trading of renewable energy credits (RECs), which quantify the replacement of high-emissions generation with lower/zero emissions alternatives, such as wind.

The monetization and liquidity of the RECs must be at a level to reasonably predict a cash stream for a project pro forma. This requires several factors in the REC trading infrastructure. First, a large enough trading pool must exist to assure enough buyers and sellers across a given geographical region. Second, the trading pool must offer some flexibility in "banking" liabilities and credits to allow for the intermittency of renewable fuel sources to garner their full value. Third, credible parties must perform regularly a uniform assessment of generating sources to quantify and certify environmental benefits.

In the coming decades, we expect a greater policy convergence (than is now evident) as the necessary framework for a robust green energy credit-trading program to fall into place. The adoption of the Kyoto Protocol and its global market for GHG emissions trading is expected to provide much of the impetus in this important area. What we learn from the trading mechanism launched by Kyoto will serve as critical lessons as RECs become liquid commodities traded across the globe.

CONCLUSION

Traditional fossil fuel sources that have helped set the stage for the industrial revolution eventually will disappear. Until that time, coal, natural gas, and oil will continue to play a vital role in our economy. Nevertheless, renewable energy sources such as wind power will begin to displace these finite fuel sources, and technology advances will continue to determine the viability and rate of adoption of new energy alternatives.

The investment required to fund this transformation will be significant, with some estimates in the trillions of dollars over the course of this century. In order to attract investment, research, or capital equipment, policy must be stable and at the same time must address barriers (such as intermittency) that exclude the promise of new fuel solutions like wind power. The traditional "free market" energy portfolio increasingly must come under pressure to consider the monetization of environmental impact and cost of risk/fuel price volatility. Ultimately, wind power and other renewable resources will indeed become the dominant fuels in the latter part of the 21st century.

Mark Little is *vice president* of GE Energy's power generation segment headquartered in Schenectady, New York. He can be contacted through Mary McCann at T: 661-823-6732 and E: Mary.McCann@ps.ge.com.

CHAPTER 9

Of Crystal Balls and Market Fundamentals

Anticipating GHG Prices

By Dr. Mark C. Trexler

The Greenhouse Gas (GHG) market has been the subject of enthusiastic speculation ever since the *Financial Times* predicted a $100 billion annual market by as early as 2010, far eclipsing any other environmental commodity market.

At one level, everyone likes the idea of a GHG market. From the standpoint of industrialized countries and regulated companies, a market for GHG reductions could dramatically reduce compliance costs associated with international targets and domestic mandates. From the standpoint of developing countries and project developers, a market for GHG reductions could deliver large quantities of capital for energy, forestry, and other projects that otherwise might not find the needed funds.

One reason everyone can see themselves as a "winner" in the coming GHG market is that we're early enough in the market's development, and the commodity we're talking about is still sufficiently undefined, so that everyone is free to believe almost anything they want to about the future market. This "Wild West" aspect of the market is reinforced by the wide range of what's being paid today for different kinds of GHG reductions, as compared to what prices might need to be paid in the future:

- The Chicago Climate Exchange touts a market-clearing price of $0.84/ton in its latest auction;
- New power plants in Oregon can "pay off" their mitigation obligations at the rate of $0.85/ton;
- The current price of "Kyoto compliant" reductions from projects in developing countries is $3–5/ton;
- Some companies in Europe have been paying $10–13/ton for credits

that will help them meet their obligations under the European Union's Emissions Trading System (E.U. ETS);

- Some companies in the United States have been paying $10/ton or more for voluntary reductions with high public-relations value;
- Many market participants or potential participants (including the Russians) have assumed that credits would be worth $10–20/ton during the 2010 timeframe; and
- Modeling by the Stanford Energy Forum of what it would take to stabilize atmospheric concentrations of CO_2 in the atmosphere suggests a market value of at least $75/ton.

Unfortunately, the "Wild West" aspect of today's early market stages confuses thinking about the GHG market's future. Some observers, trying to apply "typical" commodity thinking to the GHG arena, have concluded that the GHG market is too chaotic, that one forecast may be just as good as another, and that there is nothing to be gained from analyzing the subject of GHG prices more seriously. This approach tends to hide how important GHG price anticipation is to the development of climate change policy, both governmental and corporate, and leads to corporate decision-making that is almost by definition suboptimal. This is because many companies end up either ignoring the future cost or value of GHG emissions reductions or valuing them in ways that may not support sound corporate decision-making (see Box 1).

The bottom line is that many people approach GHG price anticipation in the same way they would approach price forecasting for a typical commodity; they come away frustrated by the perception that GHG price anticipation is an exercise in crystal-ball gazing. They don't realize that tailored GHG market analysis could significantly aid corporate and investment decision-making. The objective of this chapter is to explore the current state of GHG market forecasting and build the case for a "tailored analysis" approach that can guide corporate policy development.

WHY GHG PRICE ANTICIPATION IS KEY!

Recognizing that traditional commodity forecasting techniques are likely to be frustrated by the nascent GHG market, there is no avoiding the fact that GHG price anticipation is essential to governmental and corporate policymaking in many contexts:

Box 1

Translating GHG Prices into Corporate Strategy

Citing the uncertainties that exist with respect to future GHG markets, companies have been responding in several ways:

1. Many companies are simply not factoring future GHG costs or credits into their corporate decision-making or investment planning. In effect, these companies are currently betting (intentionally or unintentionally) that the right price forecast is $0.
2. A smaller fraction of companies are arbitrarily choosing one or more values for GHG credits that they then use in sensitivity cases for strategic and investment decision-making. Because of weaknesses in the process of generating these credit values, however, these sensitivity cases often do not have much influence on the outcomes of the decision-making process.
3. A much smaller number of companies actually build GHG costs or credit values into all of their investment decision-making, although again the values used are often quite arbitrary.

For most companies, it has proven very difficult to move from category 1 to either category 2 or 3. What often happens is that when a company is considering a proposal to more actively incorporate a GHG price curve into corporate decision-making, a senior corporate decision maker will disagree with the whole concept or with key assumptions used in developing the price forecast being proposed. This is not per se unreasonable, since considerable potential exists for disagreement on forward-looking policy scenarios and price curve assumptions. However, to the extent this outcome leaves a company in Category 1, or at best in Category 2, the likely result is suboptimal corporate decision-making that does not utilize the best available information.

- **Can society afford emissions targets?** Many domestic politicians are asking this question, arguing that the United States cannot afford to absorb the costs of an aggressive climate change strategy. This question also guided Russian debate over ratification of the Kyoto Protocol; now the Protocol is in force, it will be a key issue in the context of second commitment period targets.
- **What standards should policy makers apply to emissions trading markets?** Such standards, particularly the additionality standard as illustrated in Figure 2, will be key to future GHG prices. Setting the standards is inevitably influenced by policy makers' expectations of future prices in GHG markets.

- **Should companies invest in emissions reductions today?** Hundreds of companies in Europe, Japan, and Canada are grappling with the question of whether they need to be in the GHG project market today, with all of its uncertainties, or whether they can afford to wait a few years and simply buy GHG credits through financial markets later. Near-term price trends are key to whether these companies' decisions make them winners or losers in coming years. This is illustrated graphically in Figure 1, where a five-million ton annual CO_2 liability for the period 2008–2012 is met either by buying credits now for prepositioning, or by purchasing credits on the spot market as needed. The differential in the NPV of the two strategies ($68 million in the former vs. $268 million in the latter) illustrates that these decisions quickly become material. Successfully buying credits now for prepositioning, however, does require informed anticipation of the market.
- **Should projects sell credits today?** Project developers interested in selling emissions reductions face the opposite question. Should

Figure 1

Managing a 5 Million Ton/Yr Liability 2008-2012

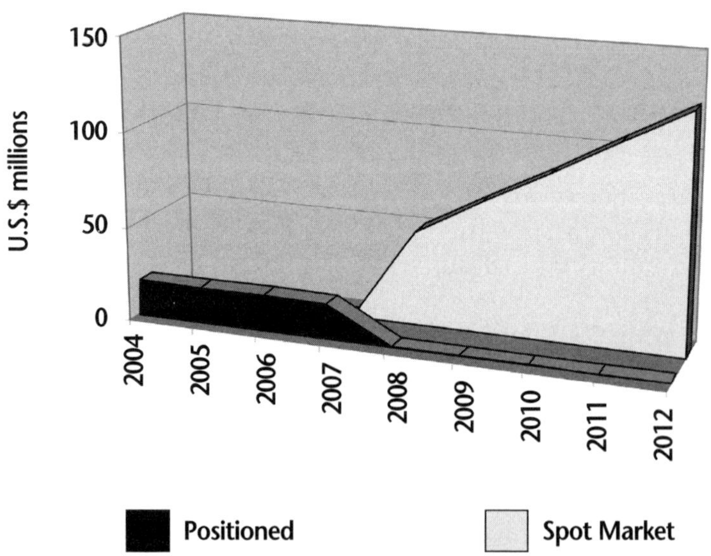

they sell those reductions today at a low price (deriving at least some benefit), or should they hold on to those reductions in the hope that prices will rise?

- **Should U.S. utilities plan for IGCC instead of pulverized coal?** It does not take a particularly high CO_2 value to swing plant economics from favoring a conventional pulverized coal plant, to favoring a coal gasification plant (IGCC). What is the rational decision for U.S. utilities today as they think about adding capacity?
- **Will GHG prices promote new technology development?** Almost everyone accepts that new technology development is indispensable to a successful long-term climate change mitigation strategy; the GHG market often is held up as a motivating driver for such technology development. But will GHG credit prices be high enough to deliver the desired outcome?

The answers to these questions are financially material to thousands of companies in the United States and worldwide. As noted in Box 1, many companies are not yet paying serious attention to these questions. For those that are in Categories 2 and 3 in Box 1, how are these questions being answered?

HOW ARE GHG PRICES CURRENTLY BEING FORECAST?

Many GHG market forecasts exist, based upon a wide variety of analytical approaches, including:

- top-down macro-economic modeling;
- bottom-up studies of energy efficiency and other technical potentials;
- extrapolating from our experience with other environmental commodity trading such as sulfur dioxide;
- projecting from historical prices in the nascent GHG market;
- estimating the future costs of technologies that will tend to cap GHG credit prices (e.g., coal-to-gas fuel switching, renewable energy technologies, and permanent disposal options); and
- "round robin" estimates, based on asking selected market participants what they think and taking an average.

These techniques generally focus on forecasting relatively near-term prices around 2010. Price forecasts range from $1 to almost $30 per ton of

CO_2-equivalent (CO_2e); the bulk of the forecasts come in at \$5–10/ton. The approaches being used to forecast GHG credit prices suffer from a number of common problems, however:

- They tend to focus on the relatively near-term future (e.g., 2010). This is a far shorter period than many companies use for investment decision-making purposes. Moreover, GHG credit prices could well change dramatically in the years after 2010;
- They tend to not specifically define the GHG commodity. A good definition requires attention to many policy and technical variables, ranging from the "additionality" and "leakage" associated with individual emissions reduction projects, to carbon-accounting protocols for forestry. Without defining a commodity that is as nebulous as GHG emissions credits, how can one realistically forecast its future price?
- They tend to suffer from the "groupthink" that can infect "round-robin" market forecasts when the number of available experts is small.

In summary, most GHG price forecasting efforts treat GHG credits as a commodity much like any other. By their very nature, the existing forecasts suggest that there is a "correct" forecast that people should use. Almost none of these forecasting techniques explicitly seek to define the GHG commodity they are talking about, a prerequisite to effective forecasting given the characteristics of the GHG market and of GHG credits.

WHAT'S SO DIFFERENT ABOUT THE GHG COMMODITY?

It is natural to assume that there must be a "good forecast" of future GHG credit prices, if only we knew where to look. Unfortunately, the uniqueness of the GHG commodity prevents this from being true. This uniqueness, in turn, results from the influence that future public policy decisions will have on both the supply and demand sides of the GHG market equation, and hence on market prices. The take-away message here is that traditional approaches to commodity price forecasting cannot be expected to lead to forecasts that satisfy the needs of corporate strategic decision-making.

Public Policy and GHG Market Demand

It is easy to see that demand in the GHG market will depend largely on policy decisions that tend to force governments and corporations to constrain their GHG emissions. Some important variables include:

- global economic growth and associated emissions growth;
- nature of future U.S. climate change policies and commitments;
- timing and severity of any post-2012 reduction targets;
- whether and how developing countries participate in global targets; and
- compliance procedures and whether countries not in compliance will be able to "simply" borrow against their future account.

It is easy to develop a scenario in which it is extremely difficult to establish international policy and in which we see only limited progress in addressing the climate-change problem. On the other hand, one can also envision a scenario in which climate change becomes a more serious political issue. The differences in global demand for GHG reductions between these two scenarios is huge—whether the U.S. is in or out of Kyoto is already a swing of some two billion tons of demand per year. Agreement on future targets could multiply GHG credit demand by more than tenfold over the next 10 to 15 years and lead to much higher credit prices than we see today.

Any view one takes regarding the future market and GHG credit prices means making important assumptions about these and other variables. They are not market fundamentals per se, and are not amenable to the same kinds of demand forecasting one might do for other commodities.

Public Policy and GHG Market Supply

What is far less intuitive than for the demand side of the GHG market is the impact of public policy decisions on the supply side of the market. A variety of policy decisions will affect the supply of GHG reductions available to meet demand at any given time, including:

- policy treatment of "hot air"[1];
- definition of "additionality" for projects in developing countries.[2] As Figure 2 illustrates, the stringency of the additionality test could dramatically affect the supply of credits available in future years;

- regulations on the treatment and accounting of forestry projects; and
- rules governing the banking of credits for future use.

As was the case with demand, any view one takes regarding the future market and GHG credit prices requires making assumptions about these and other supply-side variables. Given the variety of emissions reduction opportunities available across the six greenhouse gases covered by the Kyoto Protocol, policy decisions in these and other areas will have dramatic impacts on supply and on credit prices. This is clearly seen in Fig-

Figure 2

The Impact of Alternative Additionality Screens on the Availability and Cost of Global GHG Project-Based Reductions in 2010*

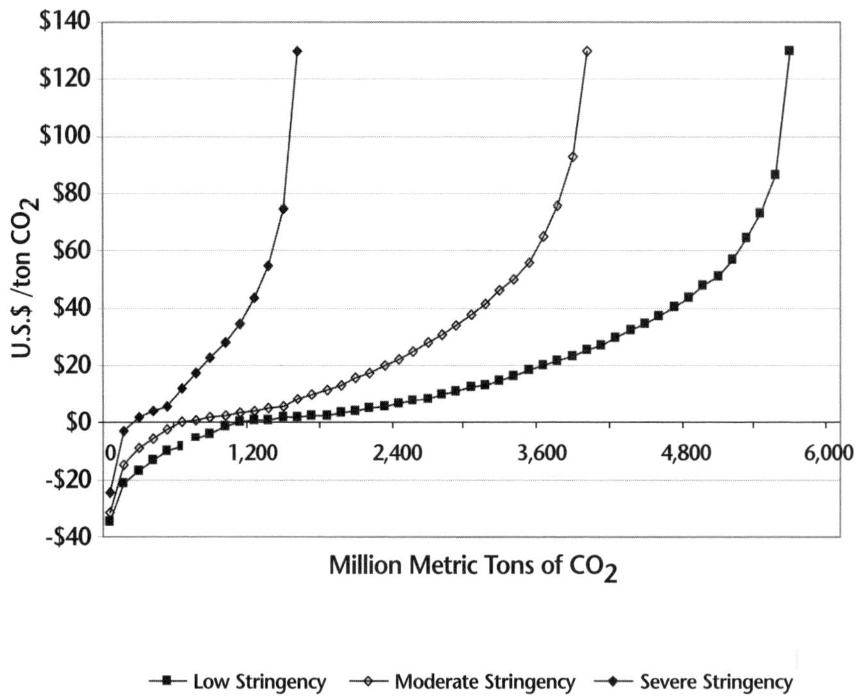

ure 2, where current supply curves illustrate how much higher prices are likely to be with stricter additionality decisions, given a set level of demand. Again, the variables discussed here are not typical commodity supply variables, and most likely are not amenable to the same kinds of supply forecasting that one might conduct for other commodities (and which is reflected in several modeling approaches often used to anticipate GHG prices).

Other Variables Likely to Influence GHG Commodity Prices

Policy variables are not the only factors that will affect supply and demand, and ultimately the price of GHG credits. Other key variables include:

- technical barriers to bringing reductions to market (e.g., project development lead times).
- responsiveness of credit sellers to market developments.
- availability of GHG project financing.
- energy prices and their implications for the cost of achieving CO_2 reductions. Table 1 illustrates how natural gas prices can dramatically affect the cost per ton of CO_2 for four representative GHG reduction technologies; thus, fossil fuel prices are an important variable for estimating future GHG credit prices.
- market psychology among buyers and sellers, based on their view of future politics and the future of the market itself.

Table 1

Cost in U.S. Dollars per Ton of CO_2 Reduced for Typical Offset Projects at Different Natural Gas Prices

	Natural Gas Price		
Typical Project	**$2.00 / MMBtu**	**$4.00 / MMBtu**	**$8.00 / MMBtu**
Coalmine Methane Capture	$5.80	$0.70	(negative)
Large-Scale Wind Energy	$45.78	$19.11	(negative)
Coal-to-Gas Fuel-Switching*	$15.12	$72.44	$187.07
Pulverized Coal CO_2 Capture**	$52.31	$59.76	$74.64

Source: Trexler Climate + Energy Services, Inc. Cli Mit Project Typicals Database©, 2003.
* Assumes coal prices stay constant.
** Lost electricity sales are assumed due to the energy penalty associated with CO_2 capture.

Conclusions Regarding the GHG Commodity

The GHG market is not yet a commodity market: the "commodity" being bought and sold now and into the future is simply not yet sufficiently defined. Moreover, there may not be just one market—there may be a market-clearing global price under the Kyoto Protocol, a different price for reductions within Europe under the E.U. ETS, and different prices for reductions occurring in response to demand drivers in the United States. What this discussion of market variables makes clear, however, is that:

- very different GHG price forecasts should accompany alternative policy scenarios;
- GHG credit prices will change dramatically over time in response to evolving public policy; and
- market psychology will be a key factor in the near term.

For companies where GHG emissions reduction mandates and the associated price of GHG credits are not material to their financial outlook, it makes sense to walk away from this issue at the current time. There are in fact significant uncertainties, and there is no "off the shelf" solution they can use. If the issue is not material to them, it probably makes sense for them to ignore it.

For more and more companies, however, the complexity of the climate change issue does not justify ignoring the issue of the future market value (or cost) of GHG reductions. The issue may be obviously material to them, as in the case of large energy companies, or key stakeholders may see the issue as material to the company (such as in the context of Sarbanes-Oxley Act disclosure requirements and new SEC rules).

FOR COMPANIES WHO MUST FORECAST

Those most affected by emerging CO_2 reduction regimes (the energy industry, sectors with high-energy intensity, or those particularly vulnerable on this issue when compared to competitors) are making business decisions that will determine their future environmental liabilities and assets. Expectations of future GHG prices (implicit or explicit) are central to ensuring that these decisions are the best they can be, even in the face of today's policy uncertainties. These companies must forecast the financial liabilities (or assets) associated with business decisions they will take in the near to medium term.

Forecasting under conditions of uncertainty is nothing new. Energy companies forecast oil and gas prices all of the time, knowing that these forecasts will not be "correct" given all of the physical and market variables that guide oil pricing. Nevertheless, these forecasts provide critical policy and strategic guidance to the companies.

What is different about forecasting the GHG market is the extent to which future policy decisions will determine GHG prices. Tax incentives and other policy decisions certainly affect oil and gas markets over time, but the situation facing the GHG market, in which policy decisions are creating the entire demand for the commodity, raise this issue to a totally new level.

The reality is that many potential policy outcomes exist, and the price forecasts associated with those outcomes may be appropriate for companies in different situations with respect to the materiality of the issue. At the end of the day, we have to ask ourselves different questions than we might with other commodities. Rather than asking ourselves what future prices will be based on anticipated supply and demand and historic price trends and volatility, companies need to reframe the question:

- What is the company's economic exposure under different policy and market scenarios?
- Can we usefully anticipate shifts in policy and market trends and outcomes, and thus develop an adaptive strategy?
- What does it make sense for the company to build into its strategic planning for future GHG prices based on its view of the future and sensitivity to alternative policy outcomes?

A company looking at the GHG market from the standpoint of covering a major potential regulatory liability may want to assume a relatively aggressive policy scenario, for example, and an associated forward price forecast that has the "downside" covered. This differs significantly from the situation facing a company looking to the GHG market as an upside for credits it plans to produce and then sell, where a relatively lax policy scenario (and conservative price curve) might be a more prudent business assumption.

Recognizing this distinguishing feature of the GHG market, and re-orienting one's forecasting focus from "the right forecast" to "the future GHG policy and associated GHG prices," is a crucial breakthrough many companies need to make in order to be able to effectively grapple with this issue and respond to growing stakeholder and regulatory directives like Sarbanes-Oxley and new SEC rules.

A "BEST AVAILABLE CORPORATE FORECAST"

Recognizing that companies need to ask what the appropriate assumption is about future GHG policy and associated GHG prices does not necessarily carry forward to the conclusion that companies can come up with useful price forecast. Indeed, the value of coming to such a realization depends on the assumption that once a company has developed a comfortable policy scenario or scenarios for its strategic planning, that this scenario can be translated into a reasonable price forecast.

Based on our experience, this can be done. Once a policy scenario is built that defines the supply and demand circumstances governing the market, it is possible to quantify supply and demand in such a way as to generate a useful forward price curve forecast. In other words, we can develop considerable insight into price outcomes and what preparations it makes sense for different companies to undertake.

What is key, in our experience, is that companies should end up with a GHG forward price curve *they* will be comfortable using for business decision-making purposes. Accomplishing this goal requires a transparent and interactive approach to the price curve forecasting process in which the client's policy and market expectations are identified and incorporated. TC+ES has developed and used this approach with a number of clients in the United States, Canada, and Japan; we term the result a "Best Available Corporate Forecast" (BACF).

Developing a Best Available Corporate Forecast for GHG market prices relies heavily on a company's own situation and thinking, since it ideally incorporates:

- materiality of the GHG issue to the company;
- the value at risk;
- timing of irreversible business decisions, investments, mergers, and acquisitions for which GHG prices are material;
- a company's perceptions of climate change policy risk; and
- whether a company is managing risk or looking for upside.

Building a Best Available Corporate Forecast should involve a number of components:

- internal education across key business units;
- diversity of business unit participation in the price curve development process;
- policy scenario building and selection;
- the application of in-depth demand, supply and market analysis to agreed-upon policy scenarios;
- tracking policy changes into the future and revising the Forecast as appropriate.

A Best Available Corporate Forecast can take a number of forms, recognizing individual companies' circumstances. The Forecast could:

- consist of a single line forecast of future prices;
- be built around a Monte Carlo type of forward price curve forecast, reflecting the company's probabilistic view of policy scenarios and market outcomes;
- simply estimate the probability that the market will be character-

ized by a given maximum or minimum price (if, for example, a major investment becomes economic or uneconomic at a particular CO_2 price).

All of these approaches potentially are legitimate choices for individual companies' circumstances; moreover, individual companies may choose to use radically different future policy scenarios as the basis for their price forecasting, and end up with very different price forecasts. These facts simply recognize that the materiality of this issue varies dramatically from one company to another and that the implications of "missing the mark" with a forecast will differ as well.

While developing a Best Available Corporate Forecast has the obvious advantage of incorporating the best available policy and market information into a company's strategic planning and investment decision making, development of a corporate forecast can have other benefits as well. The process of coming up with the BACF forces companies to ask key questions about how to view the climate change issue. It also:

- forces systematic thinking about corporate risk and benefits;
- helps avoid policy and price complacency that could hurt the company in the future;
- encourages innovative thinking by business units that otherwise might not be engaged, opening up new avenues for corporate strategy directions that could pay off for the company;
- can substantially inform corporate decision makers;
- generates cross-organization learning;
- positions companies for the market and creates a foundation for a prepositioned response strategy with significant financial benefits.

Most importantly, the Best Available Corporate Forecast is all of the things that most third-party price forecasts are not when it comes to supporting corporate decision-making. If effectively done, the Forecast will have internal corporate support and should not be as likely to be skeptically received or dismissed when presented to management for action. By helping move companies from Category 1 as described in Box 1 to Categories 2 or 3, this approach should be more successful in guiding corporate strategy than almost any third-party forecast.

CONCLUSIONS

This chapter covers a lot of ground, ranging from the GHG market to corporate strategic planning. The take aways of the chapter can be summarized as follows:

- The GHG market is not yet a commodity market.
- Traditional commodity forecasting methodologies are producing conclusions that fail to address the characteristics of the GHG market and fail to effectively guide corporate policy development.
- There is no "correct" forecast available for future GHG prices because prices will be determined by as-yet largely unspecified supply and demand variables that are being or will be negotiated.
- Notwithstanding GHG market uncertainties, thousands of companies need to forecast GHG prices for reasons that include:
  - ➢ assessing future risk;
  - ➢ choosing long-term capital investments;
  - ➢ making strategic and R&D decisions;
  - ➢ determining corporate policy and legislative positions;
  - ➢ judging the advantages of acting early;
  - ➢ protecting long-term corporate competitiveness; and
  - ➢ complying with mandates like those found in Sarbanes-Oxley and new SEC rules.
- Notwithstanding GHG market uncertainties, it is possible to forecast price ranges likely to be associated with alternative policy and market scenarios, meaning that corporations for whom this is a material issue can usefully incorporate GHG planning into their strategic thinking by answering the following:
  - ➢ Which policy and market assumptions are most reasonable to use in our strategic planning, based on our economic exposure and other sensitivities?
  - ➢ What GHG prices are associated with these assumptions?
  - ➢ What elements of developing policy are most important to track in order to usefully anticipate changing trends and reflect them in our GHG price forecast?
- This approach—TC+ES's Best Available Corporate Forecast (BACF)—has evolved from TC+ES's experience with GHG market interpretation and forecasting work over the last decade, working with both large and small companies. The GHG BACF approach builds upon our

realization that it is almost impossible for companies to develop or pursue effective climate change strategies—and to allocate any significant resources to such strategies—without incorporating a GHG forward price curve into the corporate decision-making processes.

- Notwithstanding the uncertainties, companies can settle on forward price projections that are defensible for financial and regulatory decision-making purposes and which can effectively guide corporate decision-making.

As companies come to recognize these conclusions themselves, they will find themselves much better positioned for the future of the climate change issue.

---

[1] "Hot air" refers to the gap between certain countries' emissions quotas and projected actual emissions in the Protocol's first commitment period. The collapse of the Russian and Eastern European economies since 1990 means that their GHG emissions will be nearly 30 percent below their targets. This will allow those countries to sell unused quotas for a profit, even though no actual emissions reductions have taken place relative to business as usual. The role that hot air actually plays in the international market will depend on these countries' willingness to sell those allowances into the market at prevailing market prices, as opposed to holding them as a potentially appreciating asset. Thus, hot air is considered most appropriately part of the expected supply of reductions into the market, rather than reducing market demand.

[2] In a nutshell, additionality is a test designed to ensure that projects seeking to generate GHG credits are not already "business as usual," a key requirement if reductions are brought from countries without emissions caps into countries with emissions caps. To ignore the additionality of such reductions would undercut the environmental objective of emissions reduction targets.

Dr. Mark C. Trexler is *founder* and *president* of Trexler Climate + Energy Services, Inc. He can be contacted at T: (503) 231-2727 and E: mtrexler@climateservices.com.

CHAPTER 10

Integrating Greenhouse Gas Emissions Management into Capital Projects Planning

By Arthur Lee, John J. Cain, Jonathan R. Grant, Michael P. Milliet, and John H. Shinn

Issues of international interests, economics, energy technology, and public perception drive the complexity of the climate-change issue for businesses and industries. To drive business process changes at the individual company level requires integrating the assessment of greenhouse gas (GHG) emissions. Since 2002, ChevronTexaco has compiled comprehensive annual emissions inventories of its worldwide operations and interests using the SANGEA™ system. At the same time, ChevronTexaco made a public commitment to integrate GHG emissions management into its business planning processes. This chapter describes this process and provides a portfolio of tools for integrating GHG emissions management into the planning and review of new projects. These components include 1) a screening tool that requires minimal data and can quickly provide an estimate of the potential impact of GHG emissions on a project's economics; 2) the more rigorous ChevronTexaco Projector for detailed GHG and criteria pollutant emission calculations; 3) a mitigation options planning aid; and 4) a tool to evaluate the GHG implications of potential mergers, acquisitions, and divestitures. Integrating these tools is the capital projects GHG evaluation flowchart, which provides application guidance based on the potential significance of GHG concerns for each specific project.

The tools and flowchart are innovative applications in GHG emissions management. They allow capital project teams to quickly determine the magnitude of potential greenhouse gas concerns for their projects, and then tailor their analyses appropriately. This ensures that any capital project team, regardless of the significance of GHG emissions for its project, will devote the appropriate level of effort, at the right times, and with the necessary support tools, to evaluating the potential impacts and opportunities associated with those emissions.

CHEVRONTEXACO'S CLIMATE CHANGE ACTION PLAN

ChevronTexaco is responding to the concern about climate change with a fourfold plan of action. We are:

1. **Reducing emissions of greenhouse gases and increasing energy efficiency:** Our goal is to reduce emissions-per-unit output from operations. We inventory our emissions and use innovative technologies to continually improve the energy efficiency of our existing operations, new projects, and products. We are incorporating greenhouse gas emission assessments into our capital project evaluations.
2. **Investing in research, development, and improved technology:** We invest in research to improve understanding of global climate change, to identify mitigation strategies, and to improve the cost effectiveness of mitigation technology. We develop and apply cost-effective technologies that reduce the carbon emissions of producing, delivering, and consuming our products.
3. **Pursuing business opportunities in promising innovative energy technologies:** Our research and business units are actively evaluating and investing in advanced energy technologies that have the potential of being commercially viable and beneficial to the environment.
4. **Supporting flexible and economically sound policies and mechanisms that protect the environment:** We respect the varied views of partner nations on this complex issue. We assist government policy development and decision-making on energy issues and participate constructively in dialogue with a broad range of stakeholders on this complex challenge. We support the development and use of international mechanisms such as Emissions Trading, Clean Development Mechanism and Joint Implementation, which provide flexible, market-based, economically sound means to reduce emissions.

Consistent with this action plan, ChevronTexaco policy requires that GHG issues be incorporated into capital project planning. Identifying and evaluating potential GHG issues up front, before a facility is constructed, allows concerns to be addressed during project design, to minimize potential liabilities and realize potential benefits. Understanding possible impacts on the economics of capital projects (e.g., one measure would be the effect on net present value, or NPV) also allows project teams to evaluate and choose among competing projects and project alternatives. For these reasons, fully considering potential GHG issues in capital projects is a key piece of ChevronTexaco's overall GHG management strategy.

CHEVRONTEXACO'S PROCESS FOR INTEGRATING GHG EMISSIONS MANAGEMENT INTO CAPITAL PROJECTS PLANNING

Capital projects in ChevronTexaco are managed according to the five-phase ChevronTexaco Project Development and Execution Process (CPDEP):

1. Identify and assess opportunities;
2. Generate and select alternative(s);
3. Develop preferred alternative(s);
4. Execute; and
5. Operate and evaluate.

Capital project teams begin their GHG analyses in Phase 1 or early Phase 2, as soon as enough information is available to allow emissions estimates. (See the GHG evaluation flowchart in Figure 1.) The screening tool allows a team to quickly develop a rough GHG emissions projection. It then guides them through an assessment that weighs the impact of other factors on GHG issues for a project. The output of this assessment categorizes the potential GHG impact on the project as "minimal," "small," "medium," or "large."

If the potential impact is "minimal," then no further GHG analysis is necessary. On the other hand, a "large" potential impact calls for additional GHG analysis, as an integral part of the project development effort and such studies should begin early in CPDEP Phase 2 while the various project alternatives are being considered. For projects with potential GHG impacts falling between these extremes, further analysis should either be performed to help select the preferred project alternative either in late CPDEP Phase 2 (for medium potential impact projects) or as a final check once the project design is nearing completion in CPDEP Phase 3 (for small potential impact projects). ChevronTexaco Projector and the Greenhouse

Gas Mitigation Tool (discussed in greater detail below) are available to assist with further determination.

At the end of CPDEP Phase 3, once the project design has been completed, the findings are documented on the project's appropriation request form and reviewed by the relevant decision-making body.

The following sections provide more details about the individual tools.

Figure 1

ChevronTexaco's Capital Projects GHG Evaluation Flowchart

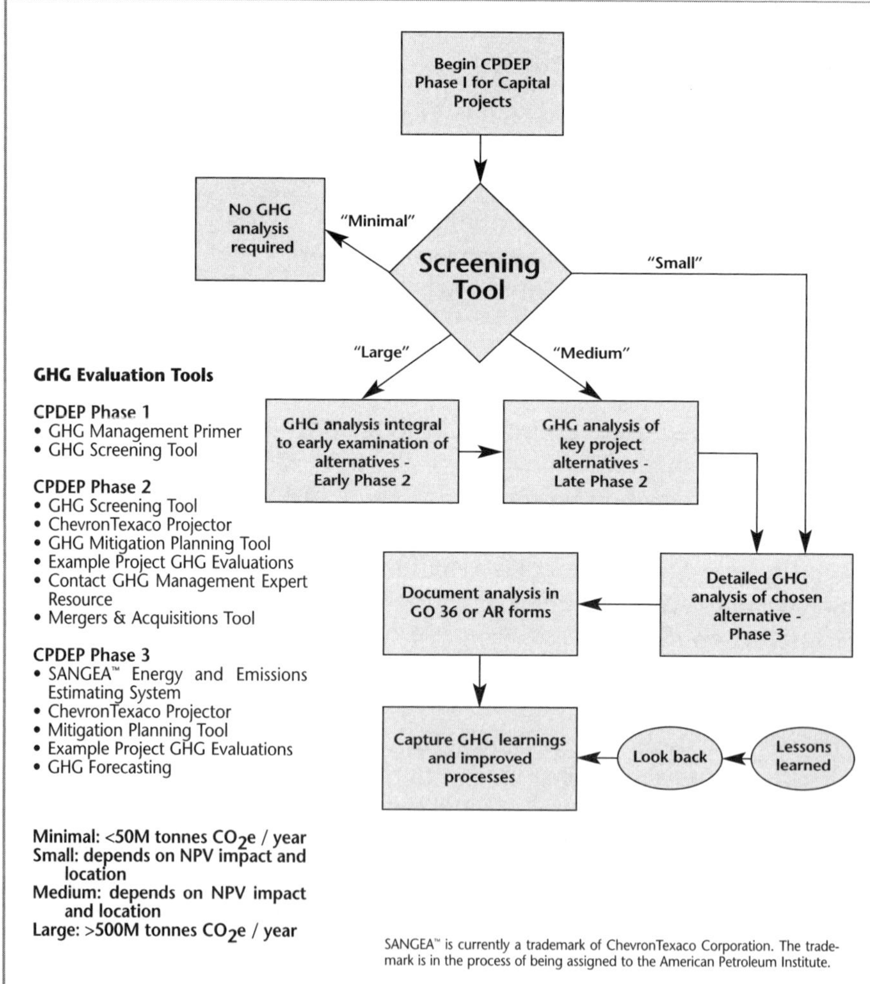

CHEVRONTEXACO'S PORTFOLIO OF GHG PLANNING TOOLS

The GHG Screening Tool

Since the range of emissions from ChevronTexaco projects varies from very large to negligible, GHG management is a key value driver for some projects and a relatively minor issue for others.

The GHG Screening Tool is used by all capital projects to conduct a preliminary screening assessment during CPDEP Phase 1 or 2. One purpose of this assessment is to quickly screen out projects without significant GHG emissions and allow them to continue additional project assessments without further GHG considerations. This preliminary appraisal is performed for all projects, regardless of size or the expected level of GHG concern.

This Screening Tool provides guidance on the order of magnitude of the GHG concerns for the project. If GHG concerns are other than negligible, the tool gives the project team direction regarding the stage at which a more detailed GHG assessment should be performed (Phase 2 or Phase 3).

As shown in Figure 2, use of the Screening Tool involves four steps. First, a project is listed on either the Exclusion List or the Inclusion List. The Exclusion List minimizes the work required for projects with historically minimal GHG emissions, such as instrumentation, routine maintenance, and R&D projects. Undertakings with total expected capital requirements of less than $5 million are also listed. In general, teams working on these kinds of projects are not expected to do rigorous GHG analysis.

However, some projects that involve GHG reductions should include rigorous analyses, even if they are listed on the Exclusion List. For these exceptions to the Exclusion List, there is an Inclusion List. This ensures that projects which reduce GHG emissions are properly analyzed and their reductions documented. For example, renewable energy projects appear on the Inclusion List. Such enterprises may provide opportunities related to carbon trading.

The next step is to estimate the GHG emissions expected from a project. The Screening Tool provides a set of order-of-magnitude emission estimating factors to allow teams to quickly assess the magnitude of emissions expected from their projects. For example:

- Burning 1 MM scfd (1 million standard cubic feet/day) of fuel gas produces about 20,000 metric tons of CO_2eq. emissions per year (1,000 normal m^3 per day produces about 750 tonnes of CO_2eq. emissions per year);

- Importing 1 MW-hr of electricity increases the project's GHG inventory by about 0.6 metric tons of CO_2eq. emissions. (Importing 1 GJ of electricity increases the project's GHG inventory by about 0.17 metric tons of CO_2eq. emissions.)[1]

While not precise, the resulting emissions estimate helps the team quickly assess the impact GHG emissions might have and provides direction regarding further evaluation.

Figure 2

ChevronTexaco's GHG Screening Flowchart

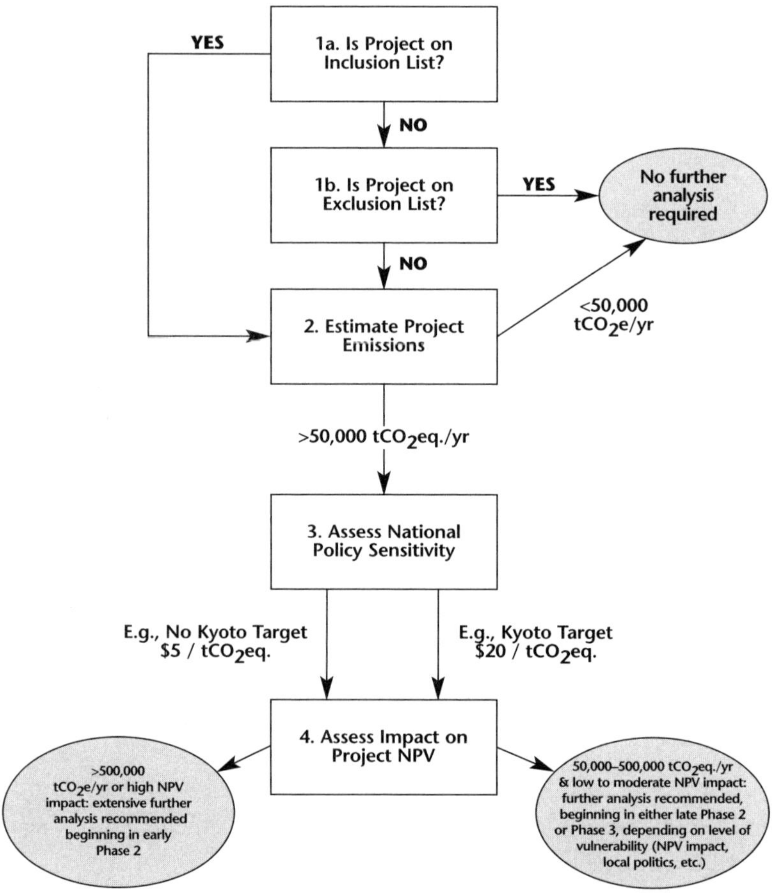

Following the order-of-magnitude emissions estimate, project teams look at the sensitivity of the host government to GHG and related issues. The Screening Tool provides guidance on the appropriate cost of carbon to use for translating the estimated GHG emissions into potential NPV impact. Countries that have ratified the Kyoto Protocol and additional specific countries (such as Australia) should be considered to have high GHG policy sensitivity, with a cost of carbon of $20/tonne. A cost of $5/tonne can be used for other countries, but in no case should the cost of carbon be lower than that. This minimum value recognizes the fact that, regardless of the level of local concern about the climate change issue, appropriate GHG management is important for ChevronTexaco. Project teams are encouraged not to rely solely on the Screening Tool but to assess the sensitivities of their host governments to determine if alternative costs of carbon (not lower than $5/tonne) are more appropriate.

Using the emissions estimate, the cost of carbon, and other project information, the project team then determines the potential NPV impact of emissions.

Finally, the Screening Tool provides an approximate indication of the significance of GHG emissions for the project. If projected emissions are greater than 500,000 tonnes CO_2eq. per year or if the potential NPV impact is relatively high, the significance of GHG for the project is classified as "large" and extensive further analysis is recommended starting in early CPDEP Phase 2 as various broad project alternatives are explored. On the other hand, if emissions are less than 50,000 tonnes CO_2eq. per year, the significance of the greenhouse gas issue for the project is classified as "minimal" and no further analysis is recommended.

If emissions fall between 50,000 and 500,000 tonnes CO_2eq. per year, then the significance is either "medium" or "small"—depending on NPV impact, host government GHG policy sensitivity, and any other considerations the project team believes are relevant. Further analysis is recommended beginning either in late CPDEP Phase 2 (as the broad project alternatives are being evaluated) or in CPDEP Phase 3 (as the preferred alternative is being developed).

ChevronTexaco Projector

ChevronTexaco Projector is an EXCEL™-based tool for rigorous GHG emissions calculations. The methodologies are consistent with the American Petroleum Institute (API) Compendium of Greenhouse Gas Emissions

Estimating Methodologies for the Oil and Gas Industry and with the SANGEA™ Energy and Emissions Estimating System, originally developed by ChevronTexaco Corporation.[2] ChevronTexaco Projector was designed specifically for capital project applications. While the SANGEA™ System has documentation and audit capabilities that facilitate its use for inventorying emissions from existing operations on a monthly basis, ChevronTexaco Projector is streamlined to more quickly evaluate multiple scenarios.

ChevronTexaco Projector has modules for calculating emissions from the various sources associated with petroleum operations, e.g., combustion, flaring, fugitives, hydrogen production, electricity, and steam import/export. There is also a miscellaneous module for users to directly input emissions not covered by any of the other modules. Additionally, ChevronTexaco Projector has a variety of features to assist project teams with various aspects of their GHG evaluation, including drop boxes for unit conversions and a sheet for NPV evaluation.

ChevronTexaco Projector results for a hypothetical scenario are illustrated in Figure 3.

Figure 3

Hypothetical Results from ChevronTexaco Projector

	RESULTS FOR PROJECT							
	CH_4	CO_2	N_2O	NO_X	CO	VOC	SOx	CO_2eq.
Combustion	5	260,863	5	463	162	9	1	262,545
Flares	56	12,754	0	7	38	25	6	14,002
Venting	0	0	0	0	0	0	0	0
Flashing Emissions	187	0	0	0	0	752	0	3,917
Transport and Storage	64	0	0	0	0	360	0	1,336
Acid Gas Removal	0	0	0	0	0	0	0	0
Coke Combustion	0	0	0	0	0	0	0	0
Fugitives	22	0	0	0	0	13	0	463
Imported Power	0	0	0	0	0	0	0	0
Exported Power	0	–13,043	0	–23	–8	0	0	–13,127
Glycol Dehydrator	9	0	0	0	0	14	0	186
Hydrogen Plant	0	0	0	0	0	0	0	0
Miscellaneous Sources	0	0	0	0	0	0	0	0
Sulfur Recovery Plants	0	0	0	0	0	0	0	0
Actual Annual Energy Usage (MMBTU/DAY)								10,901
Total Tonnes per Year	342	260,574	5	447	192	1,173	7	
CO_2eq. (Tonnes/Year)	7,179	260,574	1,569	0	0	0	0	269,322
CVX CO_2 Equivalent (Tonnes/Year)	3,590	130,287	784	0	0	0	0	134,661

GHG Mitigation Tool

The GHG Mitigation Tool helps identify and evaluate opportunities to mitigate emissions from their projects. Beyond providing information on the applicability and costs of various technologies, it outlines a general approach to assessing mitigation options for capital projects by:

1. clarifying the drivers for GHG reductions;
2. estimating the expected GHG emissions from the project, and the distribution of those emissions among individual sources;
3. reviewing potential GHG mitigation options, focusing on the largest sources;
4. considering "outside-the-fence" options, which could provide emissions offsets;
5. identifying those GHG mitigation options that are justified.

This generic approach is widely applicable across the range of projects with which ChevronTexaco is typically involved. Although the specific mitigation measures appropriate for any given project cannot be determined a priori, a number of key measures prove attractive for petroleum industry projects.

The Mitigation Tool identifies the following key measures for controlling emissions from upstream operations:

- Installing measuring devices necessary to quantify GHG emissions, especially from potentially large sources;
- Minimizing venting and flaring by capturing and using or reinjecting gas wherever it is economically feasible (if this is not possible, flaring is preferable to cold venting);
- Maximizing energy efficiency by building cogeneration of power and heat, installing energy-efficient equipment (e.g., pumps and compressors with variable speed drives), and assuring that proper maintenance and operating procedures are in place;
- Minimizing methane fugitive emissions, especially leakage from gas compressors, by installing low-leakage seals and ensuring proper inspection and maintenance procedures;
- Including flash gas separators on glycol dehydrators and using the gas as supplemental fuel in the glycol regenerator reboilers;
- Avoiding high-bleed pneumatic devices, and using instead low- or

no-bleed pneumatics, compressed air-driven pneumatics, or electronic devices.

The Mitigation Tool also identifies several key measures for controlling emissions from downstream operations:

- Maximizing energy efficiency, by
  - ➤ Installing energy-efficient equipment, such as pumps and compressors with variable speed drives, and assuring that proper maintenance and operating procedures are in place;
  - ➤ Using pinch analysis and/or other tools to optimize energy utilization; and
  - ➤ Recovering waste heat (e.g., cogeneration);
- Attempting to find an outlet other than venting for the carbon dioxide from hydrogen production;
- Minimizing methane fugitive emissions, particularly from gas pipeline systems;
- Installing the measuring devices necessary to quantify greenhouse gas emissions, especially from potentially large sources.

GHG Analysis for Mergers, Acquisitions, and Divestitures

Potential acquisitions present unique challenges to new business developers. Certain types of data, including GHG emissions data, may not be available until the due diligence stage, near the end of the acquisition process. This means that the development team would have to work with estimates in the interim.

The Mergers, Acquisitions, and Divestitures Tool provides guidance on addressing GHG considerations in potential transactions. This tool works by a) making references to order-of-magnitude estimation factors in the Screening Tool, b) recommending the use of simple ratios or other approximate means to project emissions as well as estimating historical emissions, and c) suggesting a survey of contractual and regulatory obligations for emissions limits, emissions reduction requirements, and banked emissions credits attached to the business being acquired or divested.

IMPLEMENTATION

The process and tools to support the evaluation of GHG emission impacts on capital projects were rolled out to ChevronTexaco's business units in 2003. Consistent with the ChevronTexaco Project Development and Execution Process (CPDEP) described above, the ChevronTexaco initiative that developed the process and tools discussed here has moved into Phase 5: Operate and Evaluate. Focuses in 2004 include an assessment of the overall impact of the process and its specific tools on capital project decision-making, together with an evaluation of opportunities for improving their usefulness to capital project teams.

---

1 Based on emission factors in the *Compendium of Greenhouse Gas Emissions Estimation Methodologies for the Oil and Gas Industry*, G. Harris et al. Washington, D.C.: American Petroleum Institute, 2001.

2 ChevronTexaco has made the SANGEA™ software available free of charge to the worldwide energy industry. On 22 June 2004, ChevronTexaco Corporation announced the transfer of ownership of the SANGEA™ software to the American Petroleum Institute (API) in order to continue promoting the standardization of greenhouse gas emissions accounting and reporting, and the advancement of the SANGEA™ software as the industry standard tool. The API plans to continue to make the software available free of charge to the worldwide energy industry.

Arthur Lee is *principal advisor* for Global Policy and Strategy, Health Environment and Safety Department, ChevronTexaco Corporation. **John J. Cain, Jonathan R. Grant, Michael P. Milliet** and **John H. Shinn** serve in various capacities in the ChevronTexaco Energy Technology Company. Mr. Lee can be contacted at T: (925) 842-7421 and E: rlas@chevrontexaco.com.

CHAPTER 11

Developing an Energy Efficiency and Renewable Energy Investment Fund

By John Palmisano and Deltcho Vitchev

After a long hearing within a Parliamentary Committee, a Parliamentarian once said: "Everything has been said, but not everyone has said it." Much has been said about the energy inefficiency of the transitional economies[1] and about greenhouse gas (GHG) emissions mitigation, and much has been written about the conjunction of these matters. In addition, it sometimes seems like everyone has said it. Nevertheless, this chapter goes beyond hand-wringing and lamentations and proposes a path toward practical solutions in both policy matters.

This chapter proposes new financial and policy tools to promote sustained progress in the fields of energy efficiency and renewable power while simultaneously creating substantial environmental benefits, including the reduction of GHG emissions.

The concept for a proposed Energy Efficiency and Renewable Energy Fund ("Sustainable Energy Fund") described in this chapter was developed by the authors at the request of the United Nations Economic Commission for Europe This fund should seek funding in excess of $100 million Euros and should focus on energy efficiency and renewable energy projects. The geographical focus should be on a select group of countries in the transitional economies, which have relatively high energy use combined with high energy intensity. As projects with high internal rates of return (IRR) have already been picked over in the E.U. accession countries, the geographical focus probably will narrow to Russia, Ukraine, and Southeastern Europe (Bulgaria and Romania).

Such a focused agenda aims at improving economic and environmental outcomes as well as building institutions and capacity able to promote energy security. Energy efficiency projects are further enhanced by, for example, GHG credit and quota trading. Renewable energy is additionally enhanced by tradable renewable energy credits. Together, these two environmental trading programs make environmental protection easier and cheaper. Together, the institution building and capacity building that flow from using efficient energy and renewable energy projects can generate long-lasting environmental benefits for host countries as well as their energy and environmental partners around the world. Thus, a Sustainable Energy Fund aims to build-in environmental trading indirectly and promotes capacity building, institution building, and energy security.

BACKGROUND ON ENERGY EFFICIENCY IN THE TRANSITIONAL ECONOMIES

The transitional economies, and Russia in particular, have been called the Saudi Arabia of energy efficiency. Because of cross subsidies, a surplus of cheap fuels, their focus on rapid industrialization at any cost, and social and political reasons, the transitional economies have been neither energy efficient nor focused on distributed forms of energy.

The energy inefficiency of these countries has justified many initiatives

Figure 1

Russian vs. UK Energy Intensity Rates

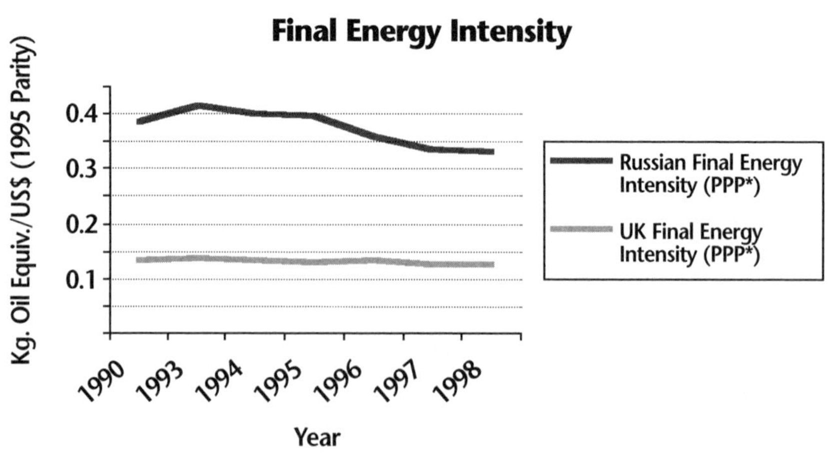

Source: World Energy Council
*PPP–Purchasing Power Parity

by donor countries and multilateral organizations. These initiatives have tried to improve inadequate host country institutions, eliminate subsidies, build human capital, transfer technologies, and encourage the development of bankable projects that will lessen the demand on energy and fuels while also reducing harmful emissions derived from fossil fuels.

The economic potential for energy savings in Central Europe is estimated to exceed 20% of the total current final consumption. In Southeast Europe and the CIS, this potential is even higher—in the range of 30–50%.[2] In Russia and Ukraine, the savings are also great. For example, the energy intensity of the United Kingdom's economy is 0.15 kg oil equivalent per 1995 U.S. dollar purchasing power parity, while it is 0.35 in Russia (see Fig. 1) and reportedly as high as 1.8 in Bulgaria (World Bank, 2003).

To meet the E.U. energy intensity levels, the transitional economies must have an ambitious work program for energy efficiency. However, resources allocated or available to energy efficiency remain insufficient to implement these objectives. The shortfall between the current situation and the targets indicates the immense commercial opportunity for an energy-efficiency fund. While energy efficiency projects offer great financial savings and high return on investment, this problem persists for many reasons.

Typical projects involve:

- Gas pipelines
- Gas compressors
- Municipal lighting
- District heating
- Residential and office buildings
- Industrial facilities
- Motors
- Boilers, and
- Instrumentation

With IRRs in excess of 20% and substantial savings in polluting fossil fuels, energy efficiency projects are attractive vehicles for achieving multiple policy objectives.

BACKGROUND ON RENEWABLE POWER

The transitional economies have vast reserves of renewable energy. Bulgaria has just about the same wind resource potential as Germany, yet Germany is a world leader in wind power and Bulgaria has virtually no wind power. Whether it is low-head hydro, coal mine, or coal-bed methane capture and use, wind, geothermal, or solar power, the region is rich in potential, but too poor to unlock these riches.

Russia is a good example of the opportunities presented by developing renewable energy resources. Russia currently uses very little of its huge renewable energy potential. In 2000, only 3% of its total primary energy supply was based on renewable energy. Of this, 2% was hydro and less than 1% all other forms.[3]

The renewable energy potential of Russia could be as high as 30% of total primary energy supply or the equivalent of more than 270 million tons of coal equivalent (MtCe). Note, however, the economic potential of renewable energy in Russia in terms of coal (millions of tons of coal equivalents) is:

- Small hydropower—65.2
- Geothermal energy—115.0
- Biomass energy—35.0
- Wind energy—10.0
- Solar energy—12.5
- Low potential heat—36.0[4]

Bulgaria, Ukraine, and Kazakhstan also have vast reserves of renewable potential. Nevertheless, unrealized potential is of no consequence. A sustainable energy fund designed to finance a critical mass of profitable projects in a few countries can create the technological push and institutional changes to move "green" energy programs out of the classroom and seminar situation and into mass production.

GREENHOUSE GASES AND CLIMATE CHANGE

Many scientists believe human activity has a profound effect on the climate of the Earth. The problem of GHG accumulation might present part of the solution. Just as GHG emissions contribute to climate change regardless of where emissions occur, climate change can be addressed by reducing GHG emissions in a different location. Thus a company in China emitting 100 tons of GHG gas emissions will have the same effect on the world's climate as a company in Russia emitting 100 tons of GHG. Because the effects are similar, a reduction of 100 tons in Russia would be able to offset 100 tons of new emissions in China. This fact suggests that emissions trading could be a useful tool in addressing climate change.

The Kyoto Protocol to the United Nations Framework Convention on Climate Change was adopted by consensus at the third session of the Conference of the Parties (COP-3) in December 1997. It contains GHG emissions targets for Annex I countries (developed countries and the transitional economies) for the post-2000 period. By arresting and reversing the upward trend in GHG emissions that started in these countries 150 years ago, the Protocol promises to move the international community closer to achieving the Convention's ultimate objective of reducing dangerous man-made interference with the climate system.

The developed countries committed themselves to reduce their collective emissions of six key GHGs by at least 5%. This group target will be achieved through cuts of 8% by Switzerland, most Central and East European states, and the European Union (the E.U. will meet its target by distributing different emission quotas to its member states); and 6% by Canada, Hungary, Japan, and Poland. Russia, New Zealand, and Ukraine are to stabilize their emissions, while Norway may increase emissions by up to 1%. (The United States had committed to reducing its emissions by 7%, but it is no longer engaged in the core features of the Kyoto Protocol process.) The six gases are to be combined in a "basket," with reductions in individual gases translated into "CO_2 equivalents" that are then added up to produce a single figure. This aggregation is calibrated based on the weighted warming potential of each GHG emission.

According to the Protocol, each country's emissions target must be achieved by the period 2008–2012. It will be calculated as an average over those five years. "Demonstrable progress" must be made by 2005. Cuts in the three most important gases—carbon dioxide (CO_2), methane (CH_4), and nitrous oxide (N_2O)—will be measured against a base year of 1990 (with exceptions for some countries with economies in transition).

Under the Kyoto Protocol, countries will have some flexibility in how they make and measure their emissions reductions. Specifically, an international "emissions trading" regime will be established, allowing industrialized countries to buy and sell emissions quotas among themselves. They will also be able to acquire "emission reduction units" by financing certain kinds of projects in other developed countries. In addition, a "clean development mechanism" will enable industrialized countries to finance emissions-reduction projects in developing countries and to receive credit for doing so. The operational guidelines for these various schemes must still be further elaborated.

HOW TRADING BENEFITS ADDS VALUE TO RENEWABLE AND EFFICIENCY PROJECTS

The analysis presented in the National Strategy Study (NSS) of Ukraine gives an overview of recent climate policy developments in Ukraine and OECD (Organisation for Economic Co-operation and Development) countries. It also includes an analysis of Ukraine's GHG mitigation potential and costs, GHG emission reduction market opportunities, information on the capacity building needs to participate in the Kyoto Protocol, GHG market options for Ukraine, project suggestions, and plan of action until 2008. The NSS plan contains recommendations almost identical to those that have been, will be, or are being done for Bulgaria, Romania, and Russia.

In support of the Ukraine NSS for managing GHG emissions, there are two scenarios forecasting economic growth and associated GHG emissions. The study clearly shows that Ukraine will have excess GHG permits under any plausible scenario at least until 2020. In terms of tons of carbon, this means that Ukraine's aggregated energy-related CO_2 emissions during the 2008–2012 commitment period of the Kyoto Protocol would be between 2,000 and 2,250 million tCO_2. Ukraine would be allowed to emit 3,360 million tCO_2 during that period (based on the 1990 level of energy-related CO_2 emissions). This means that the country will have excess emission rights of more than 1,000 $MtCO_2$, which can be sold on the international market.

The marginal abatement curve for Ukraine shows its great potential for cost-effective GHG emission abatement. Considering the period 2002–2012, 1,500 $MtCO_2$ can be reduced at costs equal to or smaller than \$10/tCO_2 and 1,000 $MtCO_2$ are in fact no-cost options. A significant portion of this potential (approximately one third of the total potential) is directly related to energy savings.

The NSS study estimated the price Ukraine could potentially obtain from these credits, given the current GHG market situation and also subject to certain market strategies. The main issues when formulating Ukraine's opportunities and strategies within the context of JI and international emissions trading of GHG quotas are the size of the GHG market, the price GHG credits may fetch on this market, and the rules of the market.

For GHG price estimates, the study uses a model provided by the World Bank. Based on this model, the study suggests a price of \$4-5/tCO_2 as a likely scenario with an expected price range spread of \$2–10/tCO_2. In addition, the study summarizes GHG prices observed in current market transactions.

These market prices are in the range of \$1–12/tCO_2. However, the model also shows that prices will drop significantly if Russia and Ukraine sell a large part of their excess AAUs on the market.

WHAT ARE THE OBSTACLES TO PROMOTING ENERGY EFFICIENCY?

Often times, "energy efficiency" seems arrested in the demonstration project mode of operation.[5] While three, four, and even ten demonstration projects contribute toward sustainable energy, institutional change is unlikely to take off at that level of activity. Furthermore, the perpetual demonstration mode of energy efficiency makes potential host country investors suspect that these approaches to energy management either cannot take root in economies in transition, or that they are flawed in some manner. While neither is the case, if these perceptions are to be overcome, it would be well to follow the maxim: "Nothing succeeds like success!"

To achieve large-scale sustainable energy and environmental outcomes within the economies in transition, advocates must scale-up their activities and find ways to attract the political and commercial interest of local political and commercial elites.

A second problem confronted by advocates of energy efficiency and renewable energy is the dominance of NGOs as recipients of funding in target countries. While part of the mainstream of Western civil society, NGOs are only now developing some influence in many of the transitional economies. This is not to suggest that NGOs have no role in promoting energy efficiency and renewable energy. Quite the contrary—NGOs are staffed by competent people and play important roles. Nevertheless, the NGO community does not have nearly the influence on the design and implementation of laws and regulations in the energy and environmental sectors in economies in transition as they do in the Western donor countries. They also lack access to sources of significant finance to fund the necessary scale-up.

To be successful, to develop and finance a pipeline of successful projects that drive policy change, institution building, and capacity building, advocates for sustainable energy must establish a new paradigm that captures the imagination of larger private sector resources and talents. Advocates must gain the attention of host country political and commercial elites on the business case that energy efficiency and renewable energy represent—high IRRs (Internal Rates of Return), technologies that work, and environmental improvements that can be monetized by trading environmental credits.

By demonstrating commercial success stories in a transparent way, insuring that these successes are replicable, and creating the policy and institutional changes that can promote replication, local elites can further develop sustainable energy projects for their own pecuniary interests. Good energy and environmental outcomes then become positive externalities from which the host countries and the world benefit.

ENERGY-EFFICIENCY FUND: CONCEPT, STRUCTURE, AND OPERATING PROCEDURES

While international organizations and donors have been able to create many demonstration projects and build substantial capacity, mass production of energy-efficiency projects has eluded funders. As a consequence, institutional change and meaningful capacity building have also eluded these institutions.

One of the rare examples of success in promoting energy efficiency and establishment of a viable pipeline of projects is the Energy Efficiency and Climate Change Mitigation Programme of the United Nations Economic Commission for Europe (UN ECE). To capitalize on the achievements of this program, it is suggested that the United Nations—along with other partner organizations and countries—develop a dedicated companion fund.[6] This fund, for convenience here called the Sustainable Energy Fund, would strive to promote the goals of the United Nations or similar organizations through a unique and trailblazing public-sector/private-sector partnership. In particular, the goals of the companion Fund will include institution building and capacity building by investing in projects that earn a commercial return. While the goal of the private-sector is to earn profits, the goal of the public-sector is often to build human capital and institutions that will systematically and in sustainable manner produce good energy and environmental outcomes.

The Sustainable Energy Fund, or Fund instruments, will provide the moneys to support the development of energy-efficiency, or renewable-energy, projects that flow from the public sector's pipeline of investment opportunities. These funds will be targeted toward a select group of countries, those that present high financial returns and who need for institution building and training.

A prudent approach would require the Fund to have a managing partner or fund manager. The managing partner could reside either within the Fund or outside of the Fund. At a later stage, legal and financial considerations will determine the legal structure and residence of the managing partner. However, in any adopted structure, the managing partner would be the core element of the Fund. Simply put, the Fund's success rests solely on the ability of the managing partner to identify, cultivate, evaluate, shape, apply the appropriate legal and financial "engineering" skills, implement, supervise and monitor projects and quantify GHG or renewable energy credits, and monetize the financial and environmental benefits that should derive from investments in energy-efficiency and renewable-energy projects. Simultaneously, public policy goals can also be served by focusing investment on those activities that reinforce each other's effects on institutions and local capacity.

Many financial instruments could be available to the fund manager. These instruments include equity, debt, loan guarantees, leasing, Build-Own-Transfer (BOT), Build-Operate-Own-Transfer (BOOT), insurance, bonds, and other instruments. Each project has its specific needs and there is rarely a one-size-fits-all solution to promote energy efficiency or renewables. In short, the fund manager needs a broad set of investment arrows in his or her quiver, and these should be used judiciously to achieve the best IRRs for the equity investors and policy goals for the public-sector investors.

The fund manager might apply resources in many ways to satisfy the Fund's financial goals and the public-sector objectives. Other tools are available and would be applied when needed and consistent with Fund goals. In our simplified model, there are three basic sources for investment—equity, debt, and credit enhancement. The private-sector is probably the best place to go to get risk capital—equity—but it is not the only place. Debt might come from both private and public sectors, while credit enhancement is likely to come from the public sector. Credits and other instruments are likely to come from the public sector or international financial institutions, while grants might come from a variety of sources.

HOW TO MEASURE SUCCESS

There is always a temptation to build and refine programs and to make them more and more elegant. This temptation should be avoided when it comes to measures of success. In the proposed partnership, the fund manager, the investors, and other stakeholders should have clear goals.

These goals can be broken down into categories:

- Policy Goals
  - ➢ Energy sustainability and energy security policy goals
    - ♦ Energy-efficient economy
    - ♦ Renewables increase in energy balance (if renewables are included within the Fund target activities)
  - ➢ Environmental improvements, specifically GHG reductions
  - ➢ Promoting market-based trading systems that derive from domestic and internal emissions trading of GHG credits
  - ➢ Building institutions and capacity in host countries
- Commercial Goals
  - ➢ Sufficient IRRs
  - ➢ Absolute returns for investors
  - ➢ Decrease the risk perception of investment in energy efficiency
  - ➢ Creating a success sufficient to attract other commercial lenders and investors to enter the market, or initiate another fund

There is a temptation to confuse performance goals and performance standards, with design goals. In general, we should reject applying design goals. The more design goals are imposed upon the policy ambitions and commercial ambitions of the fund, the less freedom the fund manager will have to produce the performance-oriented goals listed above.

In general, environmental and energy regulators have found performance goals to produce better innovation and other policy outcomes than design goals.

The Fund should have only a few, commonsense, goals. For financial investors the evaluation criteria are straightforward—sufficient IRR and positive NPV at appropriate risk. Policy goals are more difficult to define, because they often change from country to country, but at the same time these goals must be simple to measure. Furthermore, many policy planners eschew goal setting and establishing measures for success. Nevertheless, setting policy goals is imperative.

MANAGEMENT

The most important factors influencing the success of an energy-efficiency or renewable-energy fund will be the human factors—the strength of the fund manager and the ability of host country partners to supply a pipeline of bankable projects that meet the IRR and policy goals of the initiative.

"Champions" of the Fund can help promote success by working with competent NGO host country allies and by working with a fund manager that meets certain standards—in-country experience and knowledge of the technologies to be applied. However, the skills required to create the fund and those necessary to manage a successful fund are not the same. Getting money requires one set of skills; applying the money toward heterogeneous projects in a unique political and cultural setting is another.

What is needed to manage an energy-efficiency GHG-mitigation fund? The essential features of the managing partner are experience, credibility with commercial and policy people in the target countries, an ability to work well together (prima donnas can be a disaster to a well-run team), and an ability to communicate well with political and commercial leaders.

CONCLUSIONS AND RECOMMENDATIONS FOR FUTURE ACTIVITIES

Discussions with both the private and public sectors, as well as with other stakeholder groups, confirm interest in establishing a Sustainable Energy Fund able to achieve both policy and commercial goals. Clearly, there is a need for such a fund that would demonstrate that commercial profits are not incompatible with policy goals. What is now lacking is the attention and focus of the policy makers and investors to assist in establishing a flexible and commercially viable Sustainable Energy Fund.

Experience has shown that for this Sustainable Energy Fund to be successful, it must have:

- A pipeline of projects and a system for feeding that pipeline with bankable low-transaction cost projects
- A team of specialists who complement each other
- Some type of value-added proposition in terms of unique skills, proprietary tools, or special relationships

Like the weather, it seems like everyone talks about some kind of "green" fund. This might take the form of an emissions credit–buyers pool or an equity participation in renewable energy, or it might consist of loan or credit enhancement tools for energy-efficiency investment. In the future, such a fund could be used to invest the proceeds of any "carbon tax" if such is introduced. The fact that so few funds have succeeded is worth noting, and so far, money has been made by very few.

Legitimate profits can be earned and the environment protected by the proposed Sustainable Energy Fund. Energy security can be enhanced and jobs can be created. While sustainable energy has great promise, "Buyer beware" is still a good motto. Experience, skills, and knowledge of in-country problems and facilitators do not come cheap, but this human capital is worth the investment in order to achieve a cleaner and more secure future.

---

[1] For purposes of this chapter, we include under the umbrella term "transitional economies" the former Communist and centrally planned economies of Eastern Europe, Central Europe, and the Soviet Union.

[2] See IEA monograph (May 2003), "Energy Efficiency in Economies in Transition (EITS): A Policy Report."

[3] See "Renewable Energy in Russia," IEA/SLT/CERT(2003)10, International Energy Agency, March 2003.

[4] Ibid, pages 5 and 6.

[5] These arguments also generally apply to small- and medium-scale off-grid applications of renewable energy projects and many other renewable energy projects.

[6] Note the specific language. "Dedicated" means focused on a specific client, country or countries, and type or types of projects that meet certain criteria. "Companion" means aligned with, *not* within. Therefore the fund or fund-like instruments would be focused on producing outcomes consistent with the public sector's interest and not reside within a public-sector entity.

John Palmisano is a *principal* in Energy & Communications Solutions LLC, a Washington, D.C.-based project development company. He can be reached at **T**: (202) 223-2073 and **E**: john-palmisano@yahoo.com.

Deltcho Vitchev is a *director* of Renaissance Finance International, a company that specializes in structuring and implementation of projects in environmental and energy fields. He can be reached at **T**: 44 1689 861178 and **E**: deltcho@project-finance.com.

The authors provided a variety of advisory services to the United Nations Economic Commission for Europe on project development and emissions trading during the period, 1997 to 2003.

CHAPTER 12

C-Lock—A Method to Maximize Carbon Sequestration Value to Agro-forestry Producers and Purchasers

By Patrick R. Zimmerman, Karen Updegraff, William Capehart, Maribeth Price, and Lee Vierling

C-Lock is a Web-based carbon sequestration accounting and marketing tool. The C-Lock process allows agricultural producers to quantify the impact of specific land-use management practices for specific agricultural land parcels on the sequestration of carbon in soil and vegetation. It also aggregates carbon emission reduction offsets for individual land parcels into units that can be efficiently marketed. This comprehensive tool has been designed to serve as an interface to link agricultural producers, carbon sequestration science and policy, and those who wish to purchase carbon emission reduction offsets. As its most tangible output, C-Lock produces Carbon Emission Reduction Credits (CERCs) that are transparent, validated, and readily verified by an independent third party.

THE CHEYENNE RIVER SIOUX RESERVATION PILOT TRADE

In October 2004, the Institute of Atmospheric Sciences at the South Dakota School of Mines and Technology (SDSM&T) received a USDA (U.S. Department of Agriculture) Conservation Innovation Grant for its proposal "Marketing Carbon Sequestration Credits From Reduced Grazing and Conservation Practices on South Dakota Farmlands." This grant will provide funds to quantify and package credits that have been generated as a consequence of recently improved land stewardship on the Cheyenne River Sioux Reservation in South Dakota.

The Cheyenne River Sioux Tribe will market CERCs generated since 1990 due to a combination of reduced grazing, enrollment in the Conservation Reserve Program (CRP), and reduced tillage on nearly one million acres of tribal lands. The amount of CERCs to be generated through this trade is projected to be approximately 684,000 metric tons of CO_2eq, or 171,000 metric CO_2eq per annum through 2008. This projection is based on 964,000 acres registered and the mean annual sequestration rates in Table 1. These CERCs will be packaged and certified using C-Lock™ (patent pending), an online registration and modeling system developed by SDSM&T scientists.

Table 1

Projected Annual Carbon Sequestration Rates

Land Use	Carbon Dioxide Equivalent (in metric tons/acre)	Carbon (in metric tons/acre)
Land enrolled in CRP	0.74	0.22
No-till cropland	0.45	0.13
Rangeland	0.15	0.04

The unique feature of this trade will be the precise manner in which CERCs are estimated for individual land parcels, based on the specific management data supplied by landowners who register their land using the C-Lock system. Estimating sequestered carbon will not require field visits or soil sampling, although some field visits may occur as part of the validation process. The C-Lock database will track and pool marketable credits, and then prepare these credits for sale as temporary or expiring generic (e.g., vintage 2008) CERCs (see section on Marketing CERCs below).

The C-Lock system reduces transaction costs by avoiding the requirement for field sampling and third-party verification of every field. Landowners will receive payments based on their individual management inputs and site productivity, and not based on general assumptions. This is possible because the C-Lock system incorporates a secure online user interface, sophisticated databases of site-specific (climate and soil) and user data, and a numeric model to capture soil carbon dynamics. C-Lock generates annual time-step snapshots of field-level soil carbon pools based on site-specific climate and management inputs. The model-based system allows the creation of a dynamic baseline that is expected to exceed standards being developed in the U.S. Department of Energy's 1605(b) Program.[1]

By demonstrating that soil carbon trades can be packaged, certified, and verified in a transparent and cost-effective manner that maximizes potential value for all parties, this pilot trade is expected to lay the groundwork for a viable market in soil sequestration offsets. In turn, this demonstration will expand the range of market-based options available to help reduce our net national greenhouse gas (GHG) emissions over the course of the next several decades.

GREENHOUSE GASES AND THE CLIMATE CHANGE PROBLEM

In addition to oxygen and nitrogen, the air contains trace gases. Several of these trace gases, the so-called "greenhouse" gases, can affect the radiation balance of the Earth. Some greenhouse trace gases like carbon dioxide have complicated cycles. Carbon dioxide (CO_2) is released into the atmosphere as a by-product of the processing of organic matter. For example, we emit CO_2 when we breathe. The CO_2 is produced when the food we cat is respired to yield the energy we need to sustain life. CO_2 is taken out of the atmosphere by plants and organisms that use energy from the sun to convert the carbon into organic molecules like sugars, starches, and woody material. Some of the carbon that is taken up (or "sequestered") by photosynthesis is stored in long-term geological reservoirs as carbonates (i.e., limestone and chalk), as fossil fuel reserves (e.g., coal and oil), and as dissolved carbon and sediments in the ocean and soils. The carbon cycle can be thought of as analogous to the cycling of water in a hot tub. The reservoir is large compared to the amount of water in the pipes and the pump, but the circulation rates are high. Figure 1 is an illustration of our current understanding of the carbon cycle.

Figure 1

Global Carbon Cycle Schematic
(Units in Metric Gigatonnes (10^{12} MT)

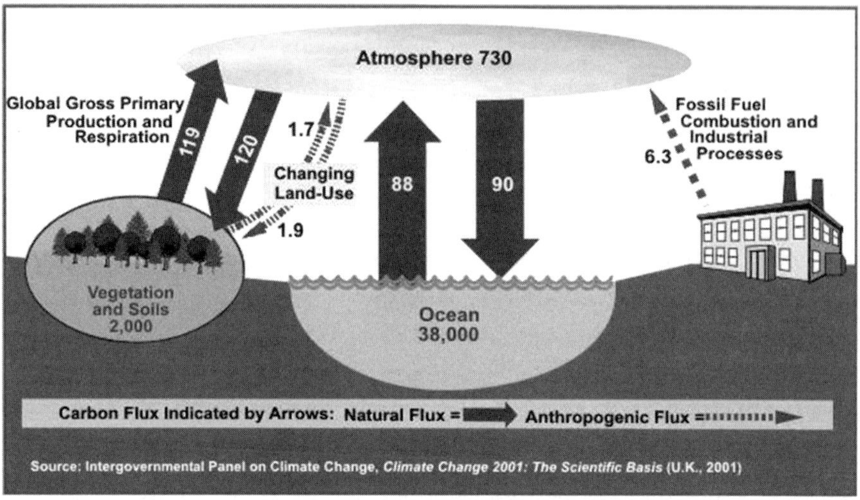

This schematic shows that the soil and ocean reservoirs of carbon are huge compared to the amount of carbon in the atmosphere as CO_2. It also shows that the flows in either direction contain huge amounts of carbon. Specifically, it shows that most of the carbon taken up by land plants and oceanic plankton (to be converted into organic matter) is re-emitted back into the atmosphere each year. Overall, an annual net increase of about 3.1 metric gigatonnes of carbon accumulate in the atmospheric reservoir as a result of fossil fuel combustion and land-use practices associated with agriculture and forestry. This represents an average annual increase in the atmospheric burden of about 0.4%. In the hot tub analogy (where the tub represents the atmosphere), this is like leaving the garden hose running at a trickle into the tub. The flow is small compared to the pumping rate, but the water level of the tub inevitably rises nonetheless.

Measurements made at remote monitoring stations around the world confirm that atmospheric concentrations of CO_2 are increasing (Figure 2). Data from stations located in latitudes where there are large temperate land masses show a large annual variation caused by the seasonal cycle of plant growth and senescence. It is notable that only about one-half of the incremental carbon dioxide emitted into the atmosphere remains as atmospheric CO_2. All of the processes responsible for sequestering the other half of the carbon emitted into the atmosphere are not completely understood. Therefore, the search to more specifically pinpoint the regions where CO_2 uptake is occurring sometimes refers to the "missing carbon sink."

Figure 2

Carbon Dioxide Trends

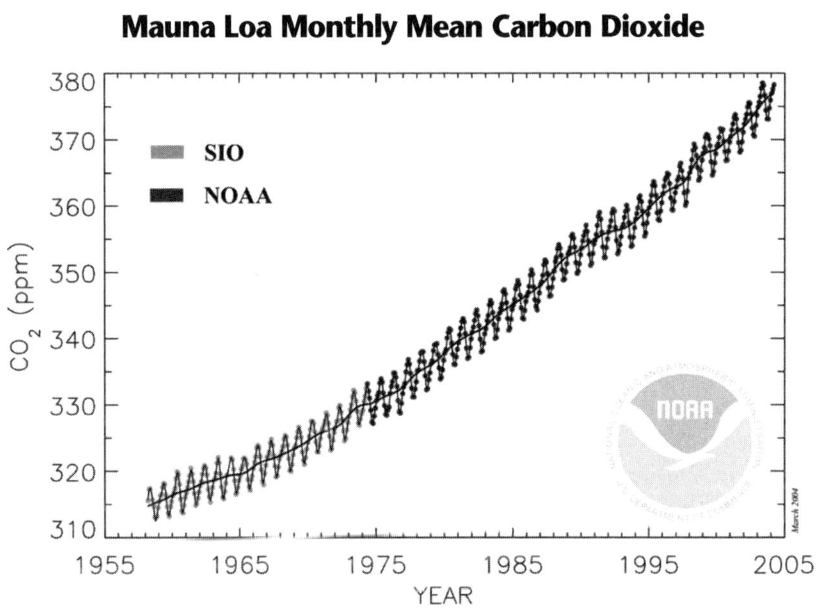

Atmospheric carbon dioxide monthly mean mixing ratios. Data prior to May 1974 are from the Scripps institution of Oceanography (SIO), data since May 1974 are from the National Oceanic and Atmospheric Administration (NOAA). A long-term trend curve is fitted to the monthly mean values. Principal investigators: Dr. Pieter Tans, NOAA CMDL Carbon Cycle Greenhouse Gases, Boulder, Colorado, (303) 497-6678, pieter.tans@noaa.gov, and Dr. Charles D. Keeling, SIO, La Jolla, California, (616) 534-6001, cdkeeling@ucsd.edu.

The changes in the atmospheric concentrations of other radiatively important greenhouse trace gases (GHG) such as methane (CH_4) and nitrous oxide (N_2O) have also been well documented. These observed changes in the atmospheric concentrations of important trace species are irrefutable. The radiative properties of the gases are also relatively well-known. The exact causes of the increased atmospheric concentrations of some greenhouse gases (such as N_2O) are somewhat uncertain, although it is certain that emissions and atmospheric concentrations of each of these radiatively active gases have increased due to human activity. In the case of CO_2, there are several independent lines of evidence that identify fossil-fuel burning as a primary cause of atmospheric buildup. In addition, the potential climate forcing due to the increases in the atmospheric concentrations have been documented, can be calculated, and are not scientifically controversial. However, the ways in which the changes in the radiation balance (caused by these GHG increases) will interact with the entire Earth's climate system are still uncertain. For example, warming of the atmosphere could lead to increases in the evaporation of water. This could increase cloud cover, which may counter the warming effect by blocking sunlight. Although the interactions and feedbacks among the composition of the atmosphere, the biology of the Earth and the physical climate are not completely understood, the consensus of scientists studying the Earth's climate system is that the net effect is likely to be an increase in the global temperature. Recently, these scientists have also agreed that the historic atmospheric temperature record probably reflects a human-induced warming outside the range of normally observed climate variability. Figure 3 shows three independently determined estimates of global surface temperature trends over the past century.

International concern that the earth's surface appears to be warming as a result of GHG accumulation led to the United Nations Framework Convention on Climate Change (UNFCCC), signed during the Earth Summit in Rio de Janeiro in June 1992 (UNEP, 1992). This treaty was ratified by the United States in October 1992. It established as its ultimate objective:

> [To achieve] stabilization of greenhouse gas concentrations in the atmosphere at a level that would prevent dangerous anthropogenic interference with the climate system. Such a level should be achieved within a time-frame sufficient to allow ecosystems to adapt naturally to climate change, to ensure that food production is not threatened and to enable economic development to proceed in a sustainable manner... (United Nations Framework Convention on Climate Change, Article 2).

In 1995, the Intergovernmental Panel on Climate Change (IPCC), established by the World Meteorological Organization and the United

Figure 3

A Comparison—Global Temperature Determined Using Surface Measurements *(top)*, Satellite Measurements *(middle)*, and European Records *(bottom)*

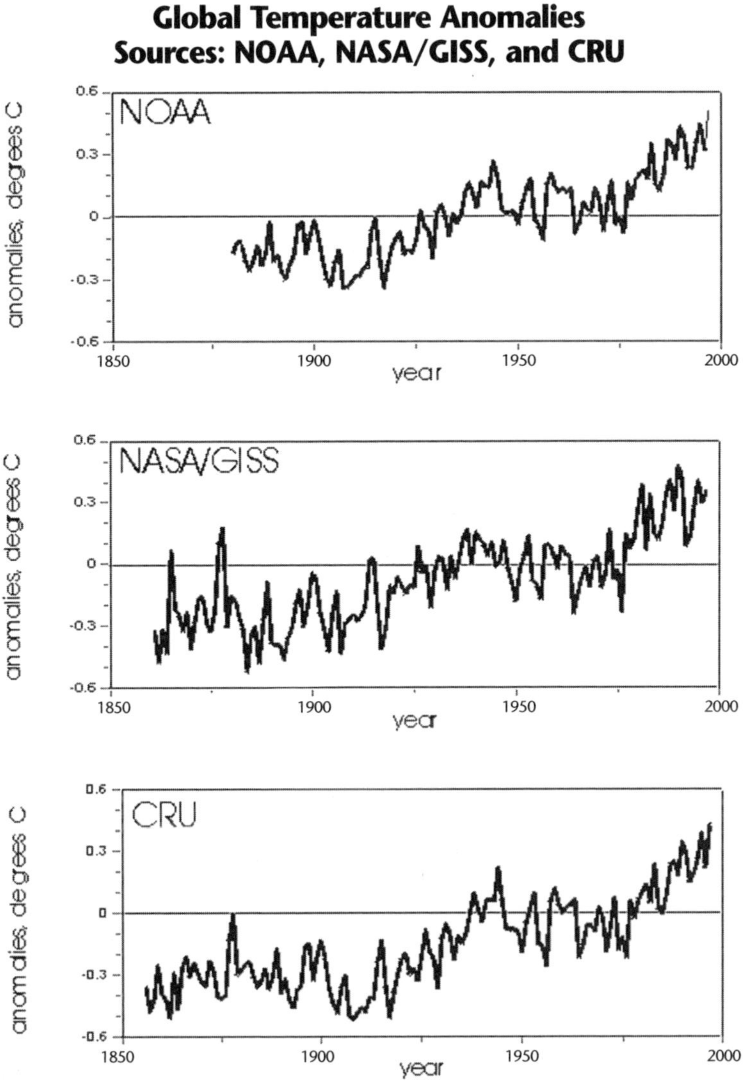

Source: A Paleo-perspective on Global Warming (http://www.ngdc.noaa.gov/paleo/globalwarming/instrumental.html)

1. NOAA: National Oceanic and Atmospheric Administration
2. NASA-GISS: National Aeronautics and Space Administration—Goddard Institute for Space Studies
3. CRU: Climate Research Unit, University of East Anglia, Norwich, UK.

Nations Environment Program, projected that, without abatement of greenhouse gas emissions:

> average global temperatures [are expected] to increase by 1.8 to 6.3 degrees F, resulting in coastal damage from rising sea levels, greater frequency of severe weather events, shifts in agricultural growing conditions from changing weather patterns, threats to human health from increased range and incidence of diseases, changes in availability of freshwater supplies, and damage to ecosystems and biodiversity... (Houghton et al., 1995).

In November 1998, the United States and 160 other countries signed the Kyoto Protocol to the UNFCCC (UNFCCC, 1998). This Protocol sets binding limits on GHG emissions for most developed countries, countries in eastern Europe, and countries in the former Soviet Union (i.e., Annex B countries). Regulated gases include carbon dioxide (CO_2), methane (CH_4), nitrous oxide (N_2O), hydrofluorocarbons and perfluorocarbons (refrigerants), and sulphur hexafluoride (SF_6). The reduction in U.S. emissions required by the Protocol is 7 percent below the 1990 emissions level, to be achieved between 2008 and 2012. No restrictions were mandated for developing countries.

The Kyoto Protocol goes into effect in 2005. Because the U.S. withdrew from the Kyoto process in 2001, various state-level initiatives are moving into the policy vacuum. As a consequence, implementation plans for trading emission allowances and offsets are rapidly evolving.

CARBON TRADING

The Kyoto Protocol contains specific provisions that allow developed countries to trade carbon credits as offsets for GHG emissions. Under the Protocol, allowable activities that provide carbon offset credit types include the development of facilities to utilize methane produced in landfills and sewage treatment facilities, the reforestation of previously cleared land, the development of biomass-based fossil fuel substitutes, the creation of renewable energy supplies (wind, hydrogen, ethanol), and the establishment of agricultural and other land-use practices that will lead to the storage of carbon in soils. No activities that sequester carbon prior to 1990 can be used to provide carbon offsets, and forest growth that would occur naturally also cannot be counted.

As a result of the belief that regulations similar to those currently included in the Kyoto Protocol will eventually be mandated in the U.S., several business ventures have been established to broker credits derived from emissions reduction or removal (sequestration) projects. For terrestrial sequestration projects, these credits are known as CERCs—defined as the removal of one metric ton of atmospheric CO_2 emissions. Many projects resulting in carbon sequestration in agricultural and forest lands have been implemented across the U.S. However, as of this date, we are not aware of any project for which a final accounting of the generation of CERCs has been completed offered for sale as a generic emissions offset in a free-market situation. (Note: Ducks Unlimited and utilities have been developing trades that include soil sequestration. However, these transactions have been exclusively one-off, with tailor-made contracts based on specific agreements between a utility and DU as an intermediary. By "free market," we mean credits auctioned in a generic way—for example, in a forum such as the Chicago Climate Exchange. Furthermore, most projects to date are associated only with general projections of carbon sequestration, not site-specific estimates, and their final accounting has not yet been completed.) It is important to note that the storage of carbon has many benefits independent of their value as CERCs. Increasing the carbon content of soils enhances the soil moisture holding capacity and improves permeability. This leads to large decreases in runoff and tends to moderate peak flows in streams after rainfall events. The result is a dramatic decrease in soil erosion. Soil erosion is the cause of the sedimentation problems occurring all along the major river drainage basins of the U.S. Increasing soil carbon stocks will lead to improvements in soil tilth, which will increase fertility and crop yields on farmland, improve the weight gain of livestock on rangeland and pasture, and drastically decrease erosion. Avoidance of sedimentation is much less expensive than remediation.

In order to maximize the sequestration of atmospheric CO_2 on agricultural and forest lands, we have designed C-Lock with the objective of minimizing transaction and verification costs while maximizing the potential value to those who purchase CERCs.

MECHANISTIC DESCRIPTION OF C-LOCK

C-Lock components include:

1. A Web page portal that provides links to informational Web sites and other resources concerning carbon sequestration science and policy.

2. A secure Web-based sign-up tool that allows agricultural producers to identify and register specific land parcels.
3. A regionally-specific geographic information system pre-populated with the climate, soil, and management variables that control carbon sequestration. This includes a geographically specific 100-year history of land use and regional agricultural practices.
4. Secure links to a regionally validated numerical model that has been designed to quantify carbon sequestration for the ecosystems of the region. The model has been validated by comparing its results with data for land where detailed agronomic measurements have been made for many years (for example, at the national network of USDA and Land-grant University Experimental Agricultural Research Stations).
5. A secure, Web-based reporting tool that lists the CERCs generated and potential CERCs.
6. An analysis of uncertainty that is used to define two pools of CERCs. One pool represents CERCs that are certified by C-Lock to exist at a specified confidence level (typically the 95 percent confidence interval, although any level could be implemented). Remaining CERCs are accumulated in the "uncertainty pool." These pools shift sizes depending on the quality of the reported data and the state of the science. The "uncertainty pool" serves as a sort of self-insurance for the certified credits.
7. The CERCs from multiple producers over multiple land parcels are in this way standardized into credits of equal value, thereby increasing their market liquidity.
8. The C-Lock process incorporates three levels of data validation:
   - Level I validation compares producer input data with lookup data for regional land use. Any discrepancies are flagged. For example, if a farmer reported corn yields of 200 bushels corn/acre in a 100-bushel region, this input would be flagged.
   - Level II verification consists of a random audit to compare satellite data for a land parcel with reported data.
   - Level III verification consists of submitting all data to a third party to operate the model and confirm the results.

In addition to these measures, regional truth-testing will be conducted as part of the U.S. Carbon Cycle Science Program. This program will use a series of tall towers and aircraft data to independently quantify the atmospheric impacts of regional carbon sequestration activities.

C-LOCK GIS DATABASES

C-Lock incorporates a geographic information system (GIS) of publicly available soil and climate databases, which are used to provide input data to a biogeochemical model, CENTURY.[2] (CENTURY was developed by scientists at Colorado State University and has been extensively validated in grassland, cropland and [to a lesser extent] forest environments. A point model, CENTURY, provides estimates of carbon storage and turnover for a specified uniform field.)

The C-Lock GIS system includes the following components:

- Soil texture and bulk density grids, which are adapted from NRCS (Natural Resources Conservation Service) digital soil survey databases. The SSURGO database[3] which provides detailed coverage, with mapping at scales from 1:12,000 to 1:63,360. SSURGO digitization has been completed in variable numbers of counties for each state, although the heavily agricultural regions are generally digitized ahead of other areas. Where SSURGO coverage is not available, the coarser STATSGO database[4] is used to fill in the gaps. Mapping scales for STATSGO are generally 1:250,000.
- Databases of mean monthly maximum/minimum temperature and total precipitation since 1900, which are created from historical and current weather data compiled by the National Climate Data Center.[5] Missing data are interpolated using zonal mean values. Each state can have up to 10 climate zones.
- Crop and farm management data which are compiled from national databases such as the USDA National Agricultural Statistics Service's historical crop statistics,[6] as well as fertilization, irrigation and management surveys conducted by the Economic Research Service. The management schedules also reflect local input solicited from expert sources such as CSREES (Cooperative State Research, Education, and Extension Service) extension personnel, extension agents, or local farmers.
- The nonproprietary biogeochemical model, CENTURY, which is driven by site-specific soils, climate, and management data. The GIS databases are used to expand its estimation capabilities to large areas.

At this time, the C-Lock team has completed the GIS databases for South Dakota and is nearing completion for Montana and Idaho.

Operationally, the farmer uses C-Lock's web interface to 1) create a private, secure account, 2) define land parcels, and 3) provide historical and current management details for each registered parcel. The farmer input data are used to (a) locate the field within the C-Lock GIS and (b) fill in management parameters.

For the years prior to 1990, the farmer selects generalized (e.g., "cropped," "grazed") management options from drop-down menus. Climate-zone-specific generic management schedules have been prepared, based on historical data for each of 5 time blocks between 1900 and 1990. From 1990 onward, the farmer is required to specify annual management details including tillage, irrigation, and fertilizer schedules. Because the simulation may run to 2030 or beyond, "future" management schedules are prepared by recycling the last 10 years of provided data unless specific alternative scenarios are provided by the registrant.

When all data entry has been completed, a "calculate" button is enabled for that field. This submits the farmer inputs to the C-Lock database, generates the appropriate weather and site parameter files, creates CENTURY management schedules, and runs the model.

C-LOCK AND STATISTICAL VARIABILITY

The C-Lock system must account for the variability in carbon accounting that results from a wide range of sources. For example, error in CENTURY predictions can result from mis-specification of site or management variables, as well as from the stochastic nature of many of its input parameters (such as weather). In order to minimize the likelihood of grossly erroneous input, C-Lock input variables are associated with quality-control limits (see above). Input data that fall outside these limits are flagged and may result in warnings or rejection of input.

Errors may also result from lack of knowledge. For example, the farmer may not have good information regarding historical management on his farm. Rather than obliging him to select an arbitrary management category, C-Lock offers a set of fallback "default" schedules. While such "default" values are not realistic for individual fields, they reflect the statistical mix of land uses within a zone and time block, as determined from county-level surveys and national agricultural data sources.

Randomness in natural processes is the most difficult to account for. For example, a specific field may not be sufficiently uniform with respect to soil type; or climate/soil data derived from the GIS may not be representative of that specific site. In order to account for these uncertainties and to estimate their impact on our reported CERCs, we have developed a Monte Carlo uncertainty estimation procedure that uses the farmer data as a basis for random value generation. These random values are used to create a specified number of weather and site parameter files, which are provided to the CENTURY model over the course of at least 200 additional runs. Input parameters that are varied include precipitation, soil texture, fertilizer or irrigation amounts, and cultivation effects.

The uncertainty estimation procedure is applied identically to the farmer's specified management schedule and to the baseline "business-as-usual" (BAU) schedule that is automatically generated by C-Lock. The BAU schedule assumes that pre-1990 management will continue for the post-1990 duration of the run. By using the same site and climate data as the farmer simulation, the parallel BAU simulation eliminates factors other than management in determining carbon stock changes.

The end result of the Monte Carlo process is two sets of carbon stock change predictions, each associated with well-defined ranges and confidence intervals. The difference between the lower confidence bound for the farmer predictions and the upper confidence bound for the BAU predictions defines the number of CERCs that can be guaranteed at the specified confidence level (usually 95 percent).

Based on test run results, the confidence associated with a C-Lock (modeled) CERC prediction is better than that associated with predictions derived from feasible rates of soil sampling. For example, based on a Tennessee study of sampling variability, a change of less than 4 metric tons of carbon/hectare would not be statistically detectable using conventional sampling methods in Midwestern soils (Conant et al., 2003). By contrast, changes of as little as 2 kg carbon/ha are theoretically detectable using the C-Lock Monte Carlo approach.

The series of carbon stock predictions generated by these runs are used to calculate distribution and confidence statistics for the CERC report. This CERC report is ultimately sent back to the farm from which the credits originated.

C-LOCK QUALITY ASSURANCE: VALIDATION AND VERIFICATION

All input data are filtered using quality control limits to trap or flag any values outside the range established for that specific region. The range of reasonably expected values is derived from and incorporated into the underlying GIS that forms a major component of the C-Lock process. This GIS contains regionally specific data for variables that control potential carbon sequestration (such as climate, soils, and prevailing crops and tillage practices). The flagging process is automated and occurs for all producer input data.

Routine audits will be applied to all clients. These audits will involve documentation of client inputs, e.g., by comparison with FSA (Farm Service Agency) records for that field. Farmers who participate in USDA commodity or conservation programs are required to provide crop and management data to the FSA. These records will eventually be available online, thereby facilitating this verification approach.

A subset of land parcels will be selected and the producer inputs will be compared with remotely sensed imagery. Satellite or airborne sensors can clearly identify and differentiate land management practices. In fact, there exists comprehensive satellite imagery extending back to approximately 1979. Satellite imagery has been used to identify land that has been tilled, land enrolled in the CRP, and many crop types. In addition, remote sensing provides an independent way to set limits on net primary productivity of a given land parcel—the starting point for carbon sequestration. We intend to automate this process in the future.

All of the data relevant to a subset of specifically identified C-Lock registered land parcels will be sent to an independent third-party auditor, along with details about how the C-Lock CERCs are quantified. This ensures that the process is completely transparent and repeatable.

The engine used to generate CERCs from the input data—a numerical model such as CENTURY—will be validated against regional soil carbon data derived from comprehensive long-term studies, where all of the variables that affect carbon sequestration have been carefully quantified and documented in the peer-reviewed scientific literature. These data are available for most U.S. states through the extensive network of regional USDA/Land-grant University Experimental Agricultural Research Stations. Measured results from land areas under long-term study (and subject to systematic sampling, monitoring, and analysis) provide higher-quality data for model calibration than could be expected from sporadic, site-specific soil sampling.

Regional verification of the impact of the widespread adoption of carbon-conserving practices will be provided via regional networks of towers designed to measure fluxes of trace gases, and other GHG monitoring stations. In particular, a comprehensive network (consisting of strategically planned facilities that include tall towers, aircraft, and flux towers) is being developed under the aegis of the North American Carbon Program (Wofsy and Harriss, 2002) to provide landscape-scale measurements of changes in the fluxes of GHG such CO_2 and CH_4. Even with a rudimentary network of stations located mostly at coastlines, scientists have been able to determine the major GHG sources and sinks. Measurements that will include vertical concentration gradients, anthropogenic tracers, specific carbon and oxygen isotopes, and fluxes of key compounds provide an independent approach to constraining flux estimates and to assessing the impact of regional mitigation practices.

The uncertainty quantification process built into C-Lock automatically ensures that certified CERCs are indeed real and associated with well-defined uncertainty limits. This process provides an automated quantitative means of discounting CERCs as a cumulative function of potential sources of error.

Actual measurements of soil carbon increments on specific land parcels are unnecessary, costly, and unable to limit uncertainty in a scientifically defensible way. However, statistically valid soil sampling protocols with appropriate sampling intervals could be designed for an aggregated region and incorporated into the C-Lock process.

The C-Lock process of emissions quantification and verification is conceptually no different than the process currently used to quantify emissions of regulated gases (such as SO_2) from specific industries. A comparison of industry and C-Lock procedures is outlined in Table 2.

Table 2

Conceptual Comparison of Emissions Quantification Procedures: C-Lock and Industry

	Industry	C-Lock
CO_2 measurements	Measurements are made within a stack	Flux measurements are made at well-documented research sites
Modeling	Base measurements are used to calibrate stack flow models*	Models are calibrated using research results and applied regionally
Validation	Extrapolated point measurements are compared with longer-term cumulative parameters (e.g. fly-ash production)	Models and data are independently verified using satellite and other monitoring data
Third-party certification	Selected data are sent to regulatory officials who verify the procedures	Data and methods are certified by an independent third party
Verification of outcome	Atmospheric data are used to independently validate the sum of the source estimates	

* Point measurements must be converted to mass flow estimates through the application of stack models. Emissions estimates must further be linked to fuel consumption using other models. Total fuel consumption can then be used to estimate total emissions for an entire facility.

MONITORING AND ENFORCEMENT OF CERC CONTRACTS

Because management practices that increase soil carbon have long-term beneficial impacts on productivity, the farmer has little incentive to cheat on his sequestration contract. However, in the event that market or other forces result in major changes, (e.g., wholesale conversion of contract fields to more intensive tillage or other uses), such changes are likely to be noticed by neighbors, in the course of FSA or NRCS site visits, or via remote sensing. The USDA has operated commodity support and conservation programs for several decades, during which it has refined methods for monitoring and ensuring program compliance. Farmers are very familiar with the process of management monitoring and verification; compliance has been excellent; and neighbors quickly expose potential cheaters.

Agricultural operations are currently unregulated with respect to air and water emissions, with the exception of concentrated animal feeding operations (CAFOs). So there is no statutory authority to enforce reductions in emissions. However, contract language can be designed to discourage defaulting by imposing financial penalties on the producer. Further, the C-Lock system promotes an approach whereby a producer commits to maintaining a pool or stock of CERCs. Self-reporting through the C-Lock interface facilitates compliance monitoring, allowing the use of simple performance contracts rather than complicated management contracts.

PERMANENCE OF SOIL CARBON

Although forms of soil organic matter can persist for over 1,000 years, soil sequestration credits are widely perceived to be impermanent because changes in management can re-emit carbon previously stored. This issue may be argued in several ways. First, the case could be made that the actions of one or two individual contract-breaking farmers in a large pool of sellers would have little impact on the total amount. C-Lock is designed to function as a credit aggregator; while defaulting farmers' CERCs may be removed from the salable pool, the total pool size (including reserve credits) should always exceed the amount sold. Therefore, the buyer does not bear the risk for contract defaulters.

The point might be further made that soil carbon does not oxidize instantly following tillage. While Midwestern soils are currently depleted of roughly 50 percent of their original carbon stocks, this depletion occurred over at least 30 years of tillage. In addition, modern tillage methods are generally much less intensive than those used in the past. Today, even in conventionally tilled fields, carbon stocks are now stable or slowly accumulating.

However, the most obvious way to address concerns about permanence is to design CERC contracts as carbon leases rather than sales. C-Lock lends itself to carbon leasing arrangements because the producer contracts to sequester and store a defined amount of carbon over the lifetime of the contract. Regular updates to the C-Lock management database facilitate low-cost monitoring and verification, and provide the producer with feedback that allows him to adapt his management strategy as needed to ensure that he meets his contract commitments.

LEAKAGE

C-Lock, as currently implemented, cannot guarantee that carbon-depleting activities will not simply be displaced onto areas outside the contract commitment. It also cannot guarantee that emissions from the farm as a whole will be reduced due to changes in management. Because farms are not regulated entities, current GHG accounting rules do not require "whole farm" accounting. For example, although no-till agriculture entails less equipment use, it typically requires heavier applications of fertilizers and agricultural chemicals, manufactured in energy-intensive processes. To address such concerns, we have developed preliminary versions of lookup tables that can be used to account for energy use in product manufacture and equipment operations. However, implementing whole-farm accounting would require significantly more information from the farmer. As the regulatory environment evolves, the modular C-Lock structure allows it to adapt.

The issue of temporal leakage is addressed through the C-Lock contracting procedure described above. The farmer commits to sequester and store (act as the steward of) a defined amount of carbon over a defined length of time. Failure to fulfill those commitments will result in the usual contractual penalties.

MARKETING CERCS

The C-Lock licensee would in effect operate as a certifier and aggregator of CERCs. CERCs available in the certified pool can be offered directly for lease to offset buyers in fixed lots, such as 100,000 $MTCO_2eq$. Lease periods of 5 to 10 years are likely to be the most attractive in an uncertain market.

In a lease-based market, offset buyers would contract with the C-Lock licensee to pay an annual fee that effectively gave it the rights to a defined tonnage of sequestered and stored carbon for the duration of the contract. Payments could be based on differential pricing for "new" (sequestered in the past year) versus "stored" (sequestered in previous years) carbon. Previously stored carbon could be leased at a lower "stewardship rate." Higher payments for "new" carbon provide incentive for the producer to continue or further enhance his carbon-conserving management (analogous to the "stewardship" and "enhancement" or "new practice" payment components in the Conservation Security Program).

The South Dakota School of Mines & Technology (SDSM&T) is in the process of establishing a for-profit corporation to fully implement the C-

Lock technology and make it available to producers. The C-Lock company will serve as a private-sector CERC certifier, aggregator, and de facto broker. The business entity will be ultimately responsible for guaranteeing the validity and long-term value of CERCs.

SUMMARY AND CONCLUSIONS

To effectively reduce our national GHG emissions and limit the potential negative impacts of rapid climate change, we need to exploit all possible mitigation options. Sequestration of CO_2 in soils and biomass is a low-cost approach to short-term mitigation, pending the availability of permanent technological solutions. One constraint to the wider acceptance of soil sequestration as a marketable emissions offset option is the cost of quantifying and monitoring credits derived from changes in land management.

C-Lock lowers the cost of generating soil sequestration offsets by:

- Allowing producers to register, estimate, and market their CERCs all in the same location, thereby reducing or eliminating the need for third-party aggregators or other middlemen.
- Reducing measurement, monitoring, and verification costs through its reliance on model-based estimation of soil carbon changes. This minimizes the need for site-specific soil sampling. It also permits better tracking of management impacts, due to the ease with which the C-Lock management database can be updated.
- Minimizing sampling requirements and transaction costs, thereby facilitating performance-based contracting, which is more equitable and economically efficient than contracts based on fixed suites of specific practices or technologies.

C-Lock leverages the best available science and provides it to producers and potential offset buyers in a practical and cost-effective manner to facilitate soil carbon credit trading. CERCs from C-Lock's certified pool are uniform units produced in a standardized, transparent, repeatable, verifiable, and cost-efficient manner. Marketed as a generic commodity, the buyer has no need to know the details of how these CERCs are generated, or to be concerned about individual CERC contracts. This is comparable to the way in which other agricultural commodities, such as corn, wheat, or beef, are marketed. At the same time, individual producers can optimize their potential income because each parcel is estimated using site-specific parameters.

The application of C-Lock in the Cheyenne River pilot trade will conclusively demonstrate that emissions offsets based on agricultural sequestration can be generated in a secure, cost-effective manner that minimizes risks for offset buyers and maximizes income for progressive land managers.

References

Conant, R. T., G. R. Smith, and K. Paustian. 2003. "Spatial variability of soil carbon in forested and cultivated sites: implications for change detection." *Journal of Environmental Quality* 32:278–286.

Houghton, J. T., L. G. Meira Filho, B. A. Callender, N. Harris, A. Kattenberg, and K. Maskell (eds.). 1995. Climate Change 1995: *The Science of Climate Change: Contribution of Working Group 1 to the Second Assessment of the Intergovernmental Panel on Climate Change.* Cambridge, UK: Cambridge University Press. 572 pp.

UNEP. 1992. United Nations Framework Convention on Climate Change. UNEP/WMO Information Unit on Climate Change. Geneva, Switzerland. (http://www.unfccc.int).

UNFCCC. 1998. Kyoto Protocol to the United Nations Framework Convention on Climate Change. FCCC/CP/1997/L.L/Add.1, 10 Dec. 1997. Geneva, Switzerland.

Wofsy, S.C. and R.C. Harriss. 2002. *The North American Carbon Program (NACP).* Report of the NACP Committee of the U.S. Interagency Carbon Cycle Science Program. Washington, D.C.: U.S. Global Change Research Program.

[1] Energy Information Administration, Office of Integrated Analysis and Forecasting. Jan. 2004. Voluntary Reporting of Greenhouse Gases Summary 2002. http://www.eia.doe.gov/oiaf/1605/vrrpt/summary/index.html

[2] CENTURY—http://www.nrel.colostate.edu/projects/century/nrel1.htm

[3] SSURGO database—http://www.ncgc.nrcs.usda.gov/products/datasets/ssurgo

[4] STATSGO database—http://www.ncgc.nrcs.usda.gov/products/datasets/statsgo

[5] National Climate Data Center—http://www.ncdc.noaa.gov

[6] USDA National Agricultural Statistics Service—http://www.usda.gov/nass/

Patrick R. Zimmerman, Ph.D., is *director* of the Institute of Atmospheric Sciences at the South Dakota School of Mines and Technology (Rapid City, South Dakota). Karen Updegraff, William Capehart, and Maribeth Price are researchers at the Institute, and Lee Vierling is a researcher at the Range Sciences Department of the University of Idaho (Moscow, Idaho). Dr. Zimmerman can be contacted at T: (605) 394-2291 and E: zimmer@ias.sdsmt.edu.

CHAPTER 13

Attracting Institutional Investment into the Australian Forestry Sector

By David G. Brand

This chapter reviews the nature of institutional investment in forestry worldwide, and how institutional investors perceive the forestry sector. It also reviews current investments and examines factors such as return requirements, liquidity, risk management, and portfolio development considerations. Considering the status of the Australian forestry sector, recommendations are made on mechanisms to encourage further investment in forestry. In particular the role of environmental markets, timber markets, and innovations in the structure of investments are examined.

INTRODUCTION

Reforestation has been described as one of the most capital-intensive activities in the world because it can require decades of patience before investors receive their returns. It has been little wonder therefore that most investment in reforestation has been by governments or the timber industry, each of whom have a direct interest in expanding timber production in rural areas. However, as existing plantations mature, both governments and industry are under increasing pressure to move forestry assets off their balance sheets, and there is a trend toward private investment in this area. This trend is well underway in the United States but is not as developed in Australia.

Australia is an attractive place to grow trees, with plentiful land, a good climate in coastal regions, and a long-term stable economic and legal system. These factors should allow Australian forestry to be attractive to domestic investors as well as competitive for international investment. Initial indications based on U.S. pension fund investments up till late 2004 (such as the Hancock Victorian Plantations and Grantham Mayo Van Otterloo [GMO] of Boston joint venture with the State Government of Tasmania) are that sound returns can be produced for forestry investors in Australia. However, available assets are currently limited, and investment products are needed with the investment parameters that institutions are seeking.

This chapter will examine the nature of institutional investment in forestry, look at some of the expectations of investors in the sector, and consider the particular characteristics and opportunities in the Australian forestry sector. The focus will be on plantation forestry and some of the emerging opportunities for environmental markets to add value beyond the traditional timber returns.

INSTITUTIONAL INVESTMENT IN THE FORESTRY SECTOR

The total pool of institutional investment in Australia today is at about U.S.$500 billion and growing rapidly. The Reserve Bank of Australia has estimated that total funds invested by institutions will rise to U.S.$1.7 trillion by 2015.[1] About 70 percent of those funds are in private superannuation (pension) funds, and the investment allocation of these funds provides a substantial contribution to the debt and equity markets. Most funds have a portfolio allocation strategy that would include such things as fixed income securities (e.g., corporate or government bonds), public equity (e.g., shares in leading companies on the Australian Stock Exchange, or ASX), and private equity or alternative investments (e.g., ownership of property, unlisted shares, or forestry assets). While all these forms of investment are used in the forestry sector, this chapter will focus on the last category of private equity investment, which is the primary business of Hancock Natural Resource Group (Australia) (HNRGA).

The so-called alternative investments, as a class, tend to be quite diverse, and could be considered almost miscellaneous assets.[2] Most institutional investors allocate somewhere between 0 and 20 percent to alternative assets, which are generally seen as potentially higher risk, less liquid, and negatively correlated with traditional debt and equity markets. Some institutions subdivide alternative assets into subcategories such as infrastructure, property, and hedge funds. All these portfolio allocation decisions are focused on the primary goal of providing high rates of return with low volatility or year-to-year variability in returns. Ultimately what every investor would like, but never achieves, is a high rate of return, year in, year out.

With growing sophistication in portfolio allocation, there appears to be increasing rigor in analysis of various assets and their characteristics. Hancock

Natural Resource Group now tracks a range of measures related to forestry assets, such as the risk-return profile, correlation with other assets, and relative contributions of cash flow and capital gains to returns (see www.hancocktimber.com for further information). These factors, among others, allow investors to analyse how a portfolio comprised of different proportions of forestry investment might perform over time.

Institutional investment in forestry has been expanding at more than 20 percent per year since 1987, and today represents approximately U.S.$15 billion. This continued steady growth in forestry investment has been due to a number of positive characteristics that flow from including forests in a portfolio. First, forests are long-term assets, and fit well with the demographic profile of many superannuation funds or insurance companies. For example, for a superannuation fund with an overabundance of members ranging from ages 40 to 50, one would expect high demands for retirement income in 15 to 25 years. Forestry is one of the few alternative assets that can be bought and held for this length of time.

Second, forests have had a negative correlation in the year-to-year variation in their returns with other major asset classes like the United States' S&P 500, corporate bonds, and real estate. At the same time, forestry assets have been positively correlated with inflation. Therefore, an investor could find that the overall volatility of their portfolio would be reduced by the addition of forestry assets. Assets that tend to increase in value when markets go down—such as gold, oil futures, and some types of hedge funds—are characterized as defensive.

The third aspect of the forestry investment is its relative risks and returns. Data from the United States over 40 years indicates that forestry investment has provided approximately 14 percent per annum returns (including inflation). This is higher than the returns from treasury bills, corporate debt, and large-cap equities (e.g., the S&P 500) but lower than small-cap equities. However, as returns rise, so does risk, and investors examine volatility versus the returns to judge whether assets are attractive. Forestry returns in the USA have tended to be low risk relative to their return rate. Another way to look at returns from forestry is to consider the discount rates that are used in buying and selling forestry assets. These tend to be in the 8 to 10 percent real (e.g., net of inflation) range—similar to projects like toll roads, parking lots, or airports. The main challenge to forestry investors is the low liquidity, and ability to accept that investments have a long-term horizon. If investors are able to accept a 10-year horizon, for example, forestry provides substantial investment benefits.

INSTITUTIONAL INVESTMENT IN AUSTRALIA'S FORESTRY SECTOR

Australia is a relatively small player in the global forestry business, with a total national harvest of about 25 million cubic meters of timber per annum (compared to approximately 190 million cubic meters/annum in Canada and 300 million cubic meters/annum in the USA). The Australian government realized that it needed to expand domestic timber supplies in the postwar era, and in the 1960s and 1970s, each State funded the establishment of large-scale pine plantations. While there were some limited privately owned plantations, the vast majority of forestry assets were owned and managed by State governments.

With increasing allocation of native forests to conservation objectives, there remained a need to expand plantations in Australia, and yet governments were less able or willing (with some exceptions) to make those investments. At the same time, in the mid- to late 1990s came the rise of the Managed Investment Scheme (MIS) industry, based on the peculiarities of Australian tax law allowing wealthy individuals in high margin tax brackets to deduct the cost of forestry investments off current income. Recently, there has been a convergence of improved investment models for tax-effective primary investments such as forestry, as well as a need for more plantation forestry for both pulpwood and sawlogs. The result was a substantial expansion of plantation forestry from 1995 to 2000, culminating with almost 140,000 hectares being planted in one year in 2000. This rapid expansion of retail or MIS investment has now led to over half of Australia's plantation estate being privately owned.

The ownership is skewed however, with about 85 percent of hardwood plantations owned by private investors and 73 percent of softwood plantations owned by government. The other trend is that about 90 percent of new plantations in recent years (say, from 1999 to 2003) have been hardwood, primarily *Eucalyptus globulus*.[3] Governments, it could be said, are largely sitting on a pool of assets that are increasingly linked to world-scale processing facilities operated by major forest industry corporations.

Clearly, such assets are candidates for privatization over time, and Victoria was the first state to sell its 100,000 hectares[4] of plantation assets, opening the door for the establishment of institutional investment in the forestry sector. These forests were sold to Hancock Natural Resource Group in 1998. Tasmania followed shortly thereafter, selling a 50-percent stake in its 50,000 hectares of softwood plantations to Grantham Mayo

Van Otterloo (GMO) of Boston. There is a general view that further privatizations will occur in Australia, with New South Wales and Queensland government-owned plantations being viewed as potential assets that would be attractive to institutional investment.

These mature softwood plantations are attractive to institutions as they have well-developed domestic markets and steady cash flow. Institutions will likely be the major owners of mature plantation forestry assets over time, as they have the ability to efficiently invest large blocks of cash and have return expectations that make them competitive buyers. Institutions have not generally invested in reforestation. This could be attributed to the overwhelming efficiency of the tax-effective MIS funds for reforestation projects, or to the demand for higher returns from greenfield investments with limited cash flow over long periods of time. However, the current trend is toward governments selling down mature softwood assets to institutional investors, and retail or MIS investors focusing on shorter rotation bluegum plantations to replace and augment woodchip supplies from native forests.

One emerging exception to this rule has been the recent acquisition of land banks underlying MIS projects by institutional investors. MIS investors receive their tax deductions for the forestry investment, which is an expense, but not for the capital cost of the land. The cost of leasing land, on the other hand, is tax-deductible. Therefore, a structure where institutional investors held the land and MIS investors were the lessees could segregate the risk-return profiles of the land and forestry assets. An initial transaction of this nature occurred when Zurich Capital Markets acquired 84,000 hectares of land from the administrators of APT, Ltd. While there was limited cash flow from the assets, the land was bought at a discounted value related to the number of years until the MIS projects would be terminated. More recently James Fielding Funds Management[5] as Trustee for the Australian Sustainable Investments Funds (ASIF) acquired 20,000 hectares of land from Timbercorp, and leased it back for forestry MIS projects.

EMERGENCE OF ENVIRONMENTAL MARKETS AND VALUES AS A FACTOR IN THE EVOLUTION OF THE AUSTRALIAN FORESTRY SECTOR INVESTMENT

Forests play an important role in a set of key environmental issues, including climate change, land and water degradation, and loss of biodiversity.[6] As governments and business search for policy solutions to these issues, there is an emerging recognition that market-based approaches may provide efficiency and lower cost outcomes for a given objective. A recent example of the successful use of a market-based approach to reduce environmental impact has been the U.S. Acid Rain Program designed to reduce emissions of sulfur dioxide (SO_2) and nitrogen oxides (NOx). Even before the market price for SO_2 allowances rose sharply in 2004, the eventual cost of reducing emissions was less than half of the lowest cost forecast before the start of the scheme.

A market-based approach is based on governments setting a policy objective (say, of cutting greenhouse gas emissions by 10 percent) and then allocating allowances to emit those gases to industry. Companies that can reduce emissions below their target level more cheaply would be encouraged to sell their excess allowances to companies with a higher cost of compliance. This facilitates the maximum emission reduction at the lowest cost. In the case of greenhouse gas emissions, there is also the opportunity to introduce offsets (carbon credits) into the mix; for example, carbon dioxide absorbed by reforestation projects could be registered and traded into the market.

Such approaches could not only include carbon sequestration,[7] but reduction in the spread of dry-land salinity, and conservation or enhancement of threatened species or natural vegetation types. Such plans are now expanding, including the New South Wales Greenhouse Gas Benchmarks initiative, in which Greenhouse Abatement Certificates can be registered based on the growth of forests established on previously cleared land since 1990.[8] Also, in New South Wales, a pilot salinity credit trade occurred in 2000 when downstream irrigators paid State Forests of NSW for additional transpiration by reforestation projects in areas with rising saline water tables.

As instruments like carbon credits or salinity credits begin to emerge and as value becomes transparent, investors may seek ways to create incremental returns or additional cash flows from forestry investments. The investments highlighted above by Zurich Capital Markets and ASIF are the first institutional investments seeking to integrate the value of carbon sequestration into the returns to the investor.[9] In these cases, state government legislation allows carbon sequestration rights to be registered separately from the ownership of the trees. This allows the investor in the land, for example, to also acquire the rights to any carbon sequestration benefits associated with the trees. In most cases, the landowner is best placed to control carbon accounting and long-term carbon sequestration maintenance obligations, particularly where the owners of the trees may vary from crop rotation to crop rotation.

The environmental credits tend to attach to "new forests," which have been planted since 1990, as opposed to mature forests or native forests. This creates a kind of bifurcation in the forestry asset class. Traditional timberland investment models will remain in effect in mature forests such as softwood resources, but reforestation projects will likely see an evolution in investment structures to encompass the relative interests of MIS investors and institutional investors, as well as the opportunities to add value via carbon trading, salinity credits, or even wind-farming rights. Institutional investors then, potentially in partnership with MIS investors, will play an increasing role in the reforestation sector as well as the mature softwood sector.

TOWARD THE FUTURE

Australia is a country with the capacity to expand its forestry sector, not only to reduce current wood products deficits, but to expand export opportunities in both softwood and hardwood timber. In addition, Australia can make use of its forestry sector to contribute to solutions to large scale or chronic environmental issues. As these opportunities unfold, we are likely to see substantial expansion of institutional investment in the sector, linked to the privatization of government assets, and emergence of new reforestation-type investment structures.

Australia could see institutional investment increase from current levels (approximately AU$1 billion or U.S.$0.7 billion in 2004) to the $3–5 billion range in the next five years if privatization and carbon markets come to pass nationally. All this would have substantial positive benefits for the forestry sector, for the environment, and for rural communities.

Legal disclaimer: *While this information comes from sources believed to be reliable, Hancock Natural Resource Group (Australia) does not warrant the accuracy or completeness of the information provided. Any projections, forecasts, or forward-looking statements in this material are a function of many assumptions and may in fact be wrong. This material is not an offer of investment advice or an offer to sell securities.*

---

[1] Axiss Australia. 2004. *Australia—A Global Financial Services Centre—Benchmark Report.* Sydney: Government of Australia Investment, Australia (see www.axiss.com.au).

[2] Daniel, W.A. and Blank, H.D. in "The Defensive Asset Class: A New Paradigm in Diversification," *Journal of Investing,* Summer 2002.

[3] *National Plantation Inventory.* 2004 Update. Canberra: Bureau of Resource Sciences, Australian Department of Agriculture, Fisheries and Forestry.

[4] One hectare is equal to 2.47 acres.

[5] See www.jamesfielding.com.au.

[6] See, for example, Brand, D.G. 2004. "Forest Investment and Emerging Environmental Markets." In Fusaro, P.C., and Yuen, M., eds. *GreenTrading: Commercial Opportunities for the Environment.* New York: GreenTrading, Inc., chapter 3, pp. 25-39.

[7] Carbon sequestration is the absorption and storage of atmospheric carbon dioxide in biomass by growing trees.

[8] See rule #5 under the NSW Greenhouse Benchmarks Scheme (see www.greenhousegas.nsw.gov.au/legislative_framework.htm#rules).

[9] See "Investors Twig to Carbon Credits," *Australian Financial Review,* March 8, 2004.

David G. Brand is *director* of the New Forests Program, at the Hancock Natural Resource Group (Australia). He can be contacted at T: 61 2 9884-4801 and E: dbrand@hnrg.com.au.

CHAPTER 14

Terrestrial Carbon Offsets for Industry Portfolios

By Dick Kempka and Dawn Browne

The mission of Ducks Unlimited, Inc. (DU) is to conserve, restore, and manage wetlands and associated habitats for North America's waterfowl. These habitats also benefit other wildlife and people. DU has developed partnerships with landowners, non-governmental organizations, and government agencies to restore habitat for North American waterfowl and aggregate credits associated with the carbon/GHG sequestered as a result of such restoration. With over two-thirds of U.S. land being privately owned, this work has enabled DU to act as an aggregator of privately owned carbon offset credits on the Chicago Climate Exchange and provide credits in volume to industrial entities that need to offset carbon dioxide emissions due to their activities. In setting up accounting and verification mechanisms for carbon/GHG credits, DU has broadened its work beyond traditional conservation activities to help develop the market structures for trading these credits and build investor confidence.

Worldwide, nearly a third of the annual anthropogenic (man-made) greenhouse gas (GHG) emissions can be attributed to land-use practices including deforestation. Therefore, appropriate land management and restoration could potentially contribute to significant reduction of GHG emissions and sequestration of atmospheric carbon concentrations.

In the U.S., agricultural soils are currently being managed as a modest carbon sink. Various studies point to the possibility of increasing this carbon sink up to 50 times current levels through habitat restoration and changes in land-use practices. Such measures would provide immediate GHG emissions reduction by altering farming management practices (i.e., equipment usage), intermediate and long-term carbon uptake through sequestration, and such eco-asset benefits as water quality improvement and conservation of habitat diversity.

LAND MANAGEMENT AND CLIMATE CHANGE

Evidence indicates that the Earth's climate is changing due to increasing atmospheric concentrations of greenhouse gases (GHG), chief among them being man-made carbon dioxide (CO_2) emissions from fossil fuel combustion and land clearing (see Figure 1).

Figure 1

Global Greenhouse Gas Emissions in 2000 by Gas (except Ozone) (in million metric tons [MMTs] of carbon equivalent or $MMTCO_2eq$)

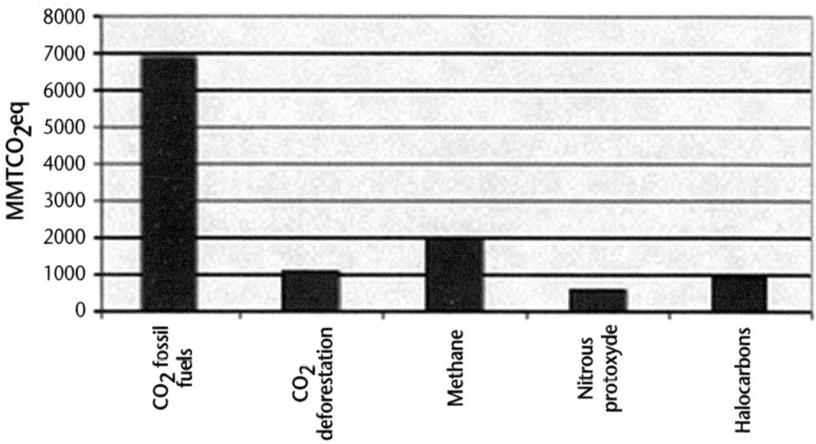

Source: IPCC, 2000

Global emissions of CO_2 from human activity have increased from an insignificant level two centuries ago to 24 billion metric tons per year in 2003. Roughly half of the anthropogenic emissions are absorbed into oceans, forests, and other natural sinks. The other half accumulates in the atmosphere, where the concentration of CO_2 is currently 379 ppm (or 33 percent above pre-industrial levels) and rising at a rate of more than 1 ppm per year (U.S. Department of Energy, 2004).

Changes over time of the distribution of land uses within a country have substantial impact on GHG emissions. Land use is characterized by the arrangements, activities, and inputs that people undertake in a specific land cover type to produce, change, or maintain it (FAO 2000). Deforestation, a specific kind of land use change, accounts for 10 to 30 percent of the annual anthropogenic GHG emissions, according to estimates in the Third Assessment Report (2001) of the International Panel on Climate Change (IPCC) (see Figure 2). In particular, oxidation of soil carbon is believed to be responsible for approximately one-third of the current 1.5 billion metric tons of carbon emitted per year globally to the atmosphere due to changes in tropical land use.

Figure 2

Sources of Carbon Dioxide Emissions

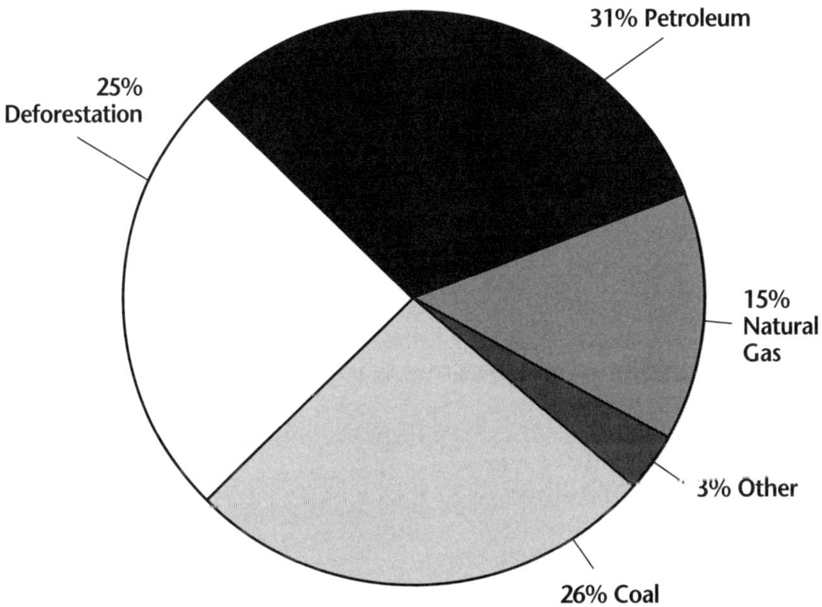

Source: IPCC, 2001

In the U.S. today, carbon emissions are not regulated. However, most companies believe there will be regulations controlling emissions within the next five years. At the state level, there are already various laws regulating carbon emissions or pending legislation. Further, international trading initiatives (such as those in Denmark, United Kingdom, and the European Union) have already begun to respond to the Kyoto Protocol that caps carbon emissions by country.

In the past two centuries, large-scale land clearing has drastically altered the U.S. landscape. Agricultural expansion and urban development have resulted in the loss or degradation of millions of acres of forests and grasslands. According to the U.S. Department of Agriculture, an estimated 40 to 60 billion metric tons of carbon may have been lost from soils as a result of forest clearing since the great agricultural expansions of the 1800s (USDA, 2004).

Land-surface disturbances influence temperature, precipitation, atmospheric circulation, and the ability of the Earth's surface to deflect solar energy. A 2004 report by the U.S. National Aeronautic Space Agency (NASA) suggests that human-caused land cover changes are at least as important an influence on climate as carbon dioxide emissions because they strongly affect regional surface temperatures, precipitation, and larger-scale atmospheric circulation.

Various carbon management methods have been proposed to mitigate the effects of emissions. They include increased efficiencies of systems for the production, conversion, and utilization of energy; alternative energy technology such as wind or solar; and emission offsets through carbon sequestration projects (geologic, ocean, and terrestrial). Given the data in Figure 2, it is reasonable to assume that terrestrial carbon sequestration should contribute at least one-quarter of the reduction of anthropogenic emissions.

TERRESTRIAL CARBON SEQUESTRATION

Terrestrial sequestration is the enhancement of the uptake of CO_2 by plants that grow on land and in fresh water, and very importantly, the enhancement of carbon storage in soils where it may reside more permanently. The U.S. Department of Energy (U.S. DOE, 2004) defines carbon sequestration in terrestrial ecosystems as the net removal of carbon dioxide (CO_2) from the atmosphere in long-lived carbon pools. These include:

- Above-ground biomass (e.g., trees, grasses)
- Long-lived products (e.g., lumber)
- Soils (e.g., organic and inorganic soil carbon)

The primary focus of terrestrial carbon sequestration is land-use and ecosystem management at the landscape or regional scale, based on the premise that this approach offers the greatest potential for enhancing carbon storage in terrestrial systems. Sequestered carbon can be quantified and measured as credits, for which monetary values can be assigned by agreement or by the marketplace.

Land and ecosystem restoration activities generate potential carbon or GHG credits in two ways. First, as farmland is converted back to native ecosystems, the new land use eliminates the emissions of carbon dioxide, nitrous oxides, and methane associated with agricultural production. Any project that removes active cropland from the landscape has an emissions-reduction benefit by removing farming practices such as fossil fuel-burning tractors, trucks that transport crops, and use of fertilizer[1] (West and Marland, 2002).

Second, land-use patterns that reduce decomposition of organic matter and increase photosynthetic carbon fixation of trees and other vegetation can achieve terrestrial carbon sequestration (see Figure 3). Reestablished vegetation captures carbon dioxide from the atmosphere and accumulates it in the plant parts and soil until saturation or net flux equilibrium is achieved. While storage periods vary, cropland converted to grassland typically takes 20 to 30 years, and reforestation takes 60 to 110 years to achieve equilibrium (Birdsey, 1996).

Figure 3

Typical Flooded Bottomland Hardwood Swamp

The degree of benefits from such restoration varies due to factors including crop type, land-use history, soil type, and the location (latitude, climate) of the property. Many current land-use management practices, such as conservation tillage or no-till farming, can increase the level of carbon in the soil and plants. Other practices that sequester carbon include converting marginal lands to wildlife habitat, restoring degraded soils, crop residue management, elimination of summer fallow, the use of winter cover crops, longer rotations, and soil erosion management.

U.S. LAND MANAGEMENT

Effective changes in management practices will require clearly defined programs that can attract the interest of landowners and be economically viable. Because 70 percent of our nation's land is in private ownership, the future of terrestrial carbon sequestration programs revolves around privately owned property.

In the Lower Mississippi Alluvial Valley (LMAV), where land restoration activities have recently begun, 87 percent of the area is privately owned and the remaining 13 percent consist of national forest and wildlife refuges, state wildlife management areas, cities, roads, and permanent open water bodies (Lower Mississippi Valley Joint Venture, 2002). Fewer than 4 million acres of the original 22 million acres of bottomland hardwoods remain, with most of this acreage in private land ownership (see Figure 4) (Hodgetts, 2000). Similar ownership patterns exist in the Great Plains of the north central U.S., where expansive tracts of grasslands have been converted to agriculture.

Recent studies indicate that U.S. agricultural soils are being managed as a modest carbon sink, accounting for net sequestration of 4 million metric tons (MMT) of carbon annually (U.S. EPA, 2003). However, many believe that these soils could be managed to store significantly more carbon. Sperow et al. (2003) estimate that U.S. croplands could be managed to sequester an additional 60 to 70 MMT of carbon per year while Lal et al. (1998) put this figure at 75 to 208 MMT. Follett et al. (2001) estimate that U.S. grazing lands could be managed to sequester an additional 29 to 110 MMT of carbon per year. These studies do not consider the option of sequestering carbon by shifting marginal croplands and grazing lands to forest. Immediate emission benefits include those listed above, while long-term benefits come from carbon storage capacity of the trees and grass that replace the farmland.

Figure 4

Forest Cover in the Mississippi Alluvial Valley Before and After Three Centuries of Drastic Land Conversion

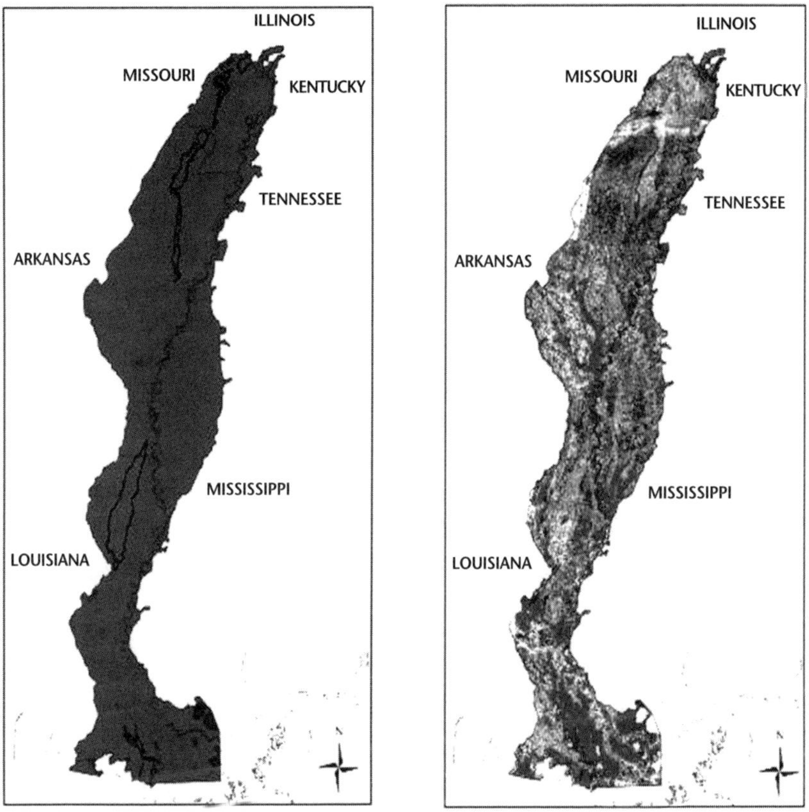

Forest cover in the U.S. South has been heavily influenced by a long history of intensive land uses. Conversion of forests and forested wetlands to primarily agricultural uses started in the late 1700s and extensive logging began after the Civil War (Wear and Greas, 2001). The graphic above depicts forest cover in the Mississippi Alluvial Valley before drastic land conversion over the last three centuries (left) and actual 2001 forest cover from Landsat ETM (Enhanced Thematic Mapper) satellite imagery.

CARBON EMISSIONS OFFSET CREDITS

Sequestered atmospheric carbon and avoided GHG emissions can be quantified and measured as credits. Projects producing offset credits must provide proof of permanent GHG reduction or otherwise account for the time that the carbon is kept out of the atmosphere. When adequately monitored and verified, the credits can be sold or traded to investors with environmental liabilities as well as to speculators.

Most corporate investors and newly formed GHG exchanges are not structured to deal directly with individual landowners who have offset credits to sell. Thus, aggregators typically secure the rights to the carbon assets derived from farmland conversion by crafting agreements with multiple private landowners. The increasing demand from industry investors drives conservation organizations, farming coalitions, and consultants to serve as carbon offset aggregators that sell carbon rights to industry partners.

In the future, the value of the carbon credits would depend on many factors including how the credits for emission reductions were obtained, subsequent land ownership, and the source of the funding used to generate the offsets. A very important factor could be governmental regulations. For example, under existing U.S. DOE guidelines, projects that use private dollars should receive full benefit and baseline protection for emission reductions. However, it is not clear whether or not projects implemented with government funding will receive full credit under the current U.S. regulatory regime.

METHODS OF PERMANENT PROTECTION

There are two straightforward methods typically used to provide land for permanent protection—perpetual conservation easements and land acquisition.

Perpetual Conservation Easement

A conservation easement is a legal agreement appended to the land deed that restricts the type and amount of development that may take place on private property. The easement holder should be a stable nonprofit 501(c)(3) organization with the resources to annually monitor and ensure protection of the property for the length of the easement. If the easement-holding organization goes out of business, then the terms of the easement may be violated without monitoring and oversight of the easement holding organization. Easements are tailored to meet the needs and interests of the landowner and easement recipient.

Conservation groups seek to protect the conservation values of the property while still providing an economic return, including carbon credits, to its owner. In order to secure the carbon rights, these groups must offer payments for easements at least as high as other available government programs, e.g., the Wetland Reserve Program (WRP). Thus, adequate easement payments are attractive to private landowners and provide incentives to convert more marginal agriculture land to its original state. After restoration, the land can still provide economic opportunities for farmers through recreational use such as hunting leases.

The length of an easement is also an important issue with regard to conservation and carbon values. Shorter-term easements, particularly those less than 30 years, are not as beneficial for preserving long-term conservation or carbon benefits. For instance, when the shorter-term easement expires, trees might be clear-cut; whereas, with a long-term carbon easement, the carbon offsets are secured by the easement holder.

Land Acquisition

A carbon-offset provider (e.g., conservation NGO) may purchase land in fee title from landowners on behalf of a corporation in need of carbon-offset credits. Typically, a conservation easement is placed on the land that protects it from development in perpetuity. This easement remains with the land even if it is sold and there are three scenarios for the future of the land:

1. The purchaser owns and manages it in perpetuity;
2. The purchaser places a perpetual conservation easement on the land and sells it to a conservation-minded buyer with restricted development rights;
3. The purchaser sells the land to a government agency required by law to protect the natural resources of the land.

In all cases, the carbon value is legally defined and protected for the buyer.

The programs described above will not only provide potential carbon credits, but will also improve wildlife habitat, water quality, and the ability of the landscape to absorb floodwaters. Additionally, carbon sequestration funding from industry will provide critical revenue to farmers and rural communities struggling with a depressed agricultural economy.

REQUIREMENTS OF QUALITY CARBON OFFSETS

In order to insure the quality of the carbon credits, offset providers must adequately address concerns of corporate investors and the marketplace. They need to provide clarity on such issues as permanence, additionality, leakage, and monitoring. The definitions of these issues under the Kyoto Protocol can be applicable across various types of transactions and regulatory regimes.

Permanence/Duration

Permanence refers to the length of time that the carbon will remain stored after having been fixed in vegetation and soil. Since protection of forests may only be temporary and tree plantations will be cut after a certain time, carbon savings achieved in forestry projects must be secured. Also, because greenhouse gases may be unintentionally released if a sink is damaged or destroyed (e.g., through forest fire or disease), it is necessary to select an appropriate carbon accounting framework for dealing with the temporal variability of sequestration. Such accounting framework should include clear definition of project duration and a timeline for project analysis. In some cases, it might include third-party insurance.

Additionality

The concept of additionality addresses the desire and recommendation that reductions of carbon must be additional to those that would have otherwise occurred under "business-as-usual" scenarios. Terrestrial carbon offset credit providers offer two types of carbon sequestration projects—(1) projects on land owned by private individuals or corporations, and (2) projects on land owned by federal, state, or local government (such as a National Wildlife Refuge, or State Wildlife Management Area, or projects funded by government subsidized programs).

While all of these offset projects are beneficial for conservation and carbon sequestration, there is debate as to whether publicly funded projects provide eligible credits since the restoration was already required without the stimulus of carbon sequestration funding. Further, in February 2002, the Bush administration announced the Clear Skies and Global Climate Change Initiatives that set a voluntary greenhouse gas intensity reduction target of 18 percent over the next 10 years. More recently, the President's FY03 budget requested a $1 billion increase in Farm Bill funding "as the first part of a ten year [2002–2011] commitment to implement and improve the conservation title of the Farm Bill, which will significantly enhance the natural storage of carbon." Specifically, the president's budget requested:

- $89 million increase for the Conservation Reserve Program (CRP)
- $800 million increase for Environmental Quality Incentives Program (EQIP)
- $176 million increase for Wetland Reserve Program (WRP)
- $16 million increase for the Forest Stewardship Program
- $254 million for a new Grassland Reserve Program

In December 2003, the USDA announced a CRP Hardwood Tree Initiative to restore up to 500,000 acres. Although the program specifies that participants will retain their rights to sell or market carbon associated with tree planting according to the guidelines, the program only appears to fund acres already authorized in the 2002 Farm Bill for purposes other than carbon sequestration. This new influx of government funding needs to be followed by clear definitions of additionality for projects funded by these various conservation programs. There will be uncertainty in the market until this issue is addressed and investors can be assured that credits purchased in association with federally funded projects will be considered additional and retain their value in the future.

Leakage

Leakage is defined as the unanticipated decrease or increase in greenhouse gas benefits outside the project's boundaries, which have occurred as a result of the project activities. A credible carbon sequestration project must reasonably demonstrate that a given land use pattern is indeed being replaced by trees or grass, without simply relocating the land-use pattern elsewhere. A project that considerably changes supply and demand can produce market effects such as reducing supply, increasing demand, or depressing the local price of wood which can cause nearby plantations to be replaced with pasture or other low-biomass land uses (IUCN, 2002).

Monitoring

Techniques are available with which to accurately and relatively easily measure or verify changes in carbon stocks. It is important to clarify the differences regarding measurement, monitoring, audit, and verification when referring to carbon stocks.

Measurement starts with the establishment and quantification of the baseline amount of carbon prior to initiation of a project. In the case of conversion of marginal agricultural land, establishment of a baseline would require documentation of crop type, irrigation practices, and other related information for the previous three to five years. Such information can come from sources such as published literature from Oakridge National Laboratory on emissions reduction rates associated with various farming practices and crop information that is available from the Farm Service Agency offices in each county in the U.S.

The amount of carbon sequestered is determined by field sampling using traditional forest and soil mensuration techniques, including measuring standing timber, estimating canopy, and chemical laboratory analysis. Recently, remote sensing technology and GIS (geographic information system) software have been used to supplement fieldwork and increase quantification accuracy across landscapes.

Monitoring involves periodic site visits to appraise seedling success and measure carbon pools to determine gains or losses from the baseline. The carbon aggregator or a consultant with the proper expertise (such as a professional forester) can perform these inspections. For example, Winrock International (a nonprofit organization offering ecosystem management services) has published *A Guide to Monitoring Carbon Storage in Forestry and Agroforestry Projects,* outlining procedures for forest carbon monitoring.

Verification is the determination that the carbon submitted for sale is actually present on the site. Audits to confirm carbon claims must be done by an independent third party with no vested interest in the results. All offset providers must demonstrate accountability, measurement, and monitoring of carbon projects to give potential investors confidence in the terrestrial carbon sequestration market and the future value of their investments.

AGGREGATORS AND OFFSET PROVIDERS

Aggregators or carbon offset providers play an important role, because the nature of trading emissions credits requires greater economies of scale than average landholders can supply. At this stage of the carbon market's development, industry purchasers and other market participants are demanding that aggregators establish clear accounting and verification mechanisms to help solidify the market structure and build investor/buyer confidence.

Conservation organizations and other entities involved in natural resources have stepped in to fill the role of aggregator of carbon-offset credits. They view land management as a vehicle to provide overall ecological gains and achieving conservation objectives while providing economic benefits to the landowner. The aggregator's role fits well with their missions since climate change clearly impacts natural resources, and this role creates potential funding opportunities for restoring natural habitat as well as improving environmental quality. In this capacity, conservation and natural resources organizations bring a uniquely rich perspective to the development and use of terrestrial carbon sequestration for emissions offset (see Table 1). Last, but not least, these private (often, not-for-profit) entities are frequently perceived by private landowners as easier to work with than government agencies because of their streamlined communication and administrative processes.

Table 1

Regulatory and Conservation Perspectives of Terrestrial Carbon Sequestration

REGULATORY PERSPECTIVE	CONSERVATION PERSPECTIVE
An emissions offset tool with ancillary environmental benefits	*A tool for offsetting emissions and restoring the sequestration capacity within the terrestrial carbon pool*
• Views the terrestrial ecosystem as storage container • Focus is on the process of "storing" offsets • Favors practices that optimize the storage process • Primary product is emission offsets from the geologic carbon pool	• Views the terrestrial ecosystem as "natural scrubber" • Focus is on the process of "restoring" an ecological function • Favors practices that benefit multiple ecological functions • Primary products are emission offsets and ecosystem restoration
Views conservation benefits as ancillary	*Views conservation benefits as intrinsic*

Source: Lower Mississippi Valley Joint Venture, 2002

An example of a nonprofit conservation organization that has taken on the role of aggregator is Ducks Unlimited, Inc. (DU). Through its habitat restoration, DU removes substantial amounts of carbon from the atmosphere, and therefore, it can function as an offset provider on the Chicago Climate Exchange. Its conservation work includes restoring wetlands and other habitats, enhancing degraded habitats, protecting endangered habitats, managing wild lands for wildlife, and influencing wildlife-friendly legislation. DU offers opportunities for energy companies to invest in its conservation projects as a way of offsetting power plant GHG emissions.

DU offers five basic types of land conversion for carbon sequestration:

- Grassland restoration in the Northern Great Plains
- Bottomland forest restoration in the Mississippi River watershed
- No-till winter cereal crops in the Northern Great Plains
- Riparian forest restoration along the East and West coasts, and
- Seasonal emergent wetlands throughout the U.S.

The net greenhouse gas flux for prairie wetlands have been assessed in recent studies by Northern Prairie Wildlife Research Center (NPWRC) of the U.S. Geological Services (USGS), Ducks Unlimited Canada, and USDA. Preliminary results indicate the potential to enhance carbon sequestration through wetland restoration in the Prairie Pothole Region. Previous work by NPWRC and the USDA suggests that prairie wetlands traditionally functioned as net sinks for atmospheric carbon but has shifted from being net sinks to net sources of atmospheric carbon as a result of cultivation, the current principal land use (Euliss et al. 2002).

To focus its conservation activities, DU uses satellite and GIS technology to identify optimal project areas within the LMAV and the Northern Great Plains. The results of this analysis are combined with information on public land ownership and existing DU restoration projects to locate appropriate private properties for carbon offset projects (see Figure 5). DU has also been working closely with the LMAV Joint Venture to expand the capabilities of the Reforestation Tracking System to track forest management prescriptions relating to carbon sequestration quantification. Recently, DU received a National Fish and Wildlife Foundation/Budweiser grant to develop a carbon tracking system that will record project location, ownership, size, land use, planting rate, site management, carbon accumulation, and many other factors associated with carbon transactions in the LMAV. Once a suitable site is identified, private land agreements that provide an unquestionable definition of carbon credit ownership are established. DU then completes the restoration work, monitors the property, and secures third-party verification of carbon sequestration based on the existing land use (e.g., crop type) and the tree species/density planted. Annual tree growth curves can then be translated into an estimated carbon value per year.

DU has contact with thousands of private landowners across the country, who value this organization's voluntary, incentive-based approach (see Figure 5). Projects with these landowners contribute to cleaner air by reducing emissions associated with traditional farming practices and by planting trees that absorb carbon dioxide. Such efforts benefit the U.S. farmer and rural communities by providing an alternate source of income during times of low commodity prices. At the same time, these efforts contribute to DU's conservation mission by increasing the amount of wildlife habitat within their priority landscapes (see Table 2).

Figure 5

Distribution of DU Habitat Projects in North America

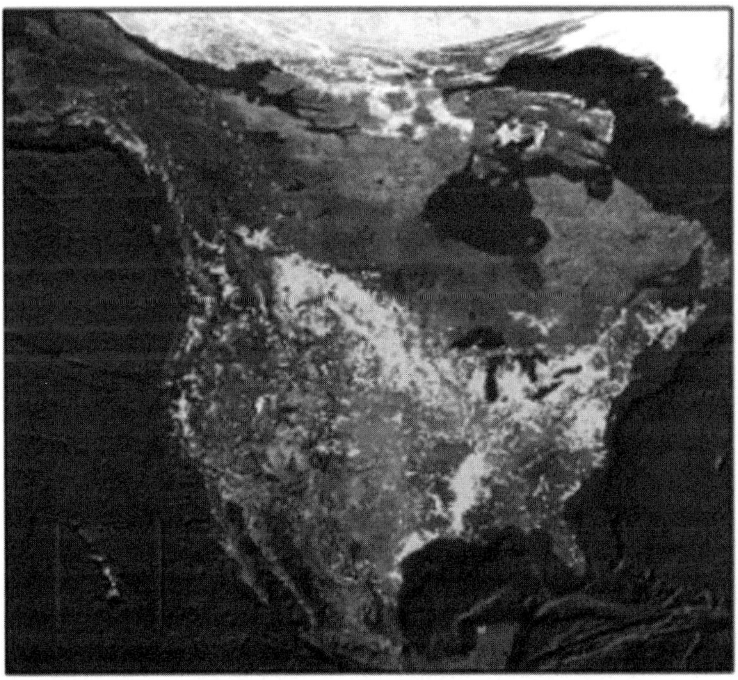

These DU partnerships demonstrate how industry and conservation can work proactively to address GHG emissions and environmental sustainability. Other conservation-based carbon sequestration approaches have been used to build successful partnerships in the Lower Mississippi Alluvial Valley among industry, non-governmental organizations (NGOs), and government agencies on public lands. However, since most remaining opportunities are on private lands, the DU model of partnership will become increasingly important.

Table 2

Duck Unlimited's Portfolio of Terrestrial Carbon Sequestration Projects*

Program Areas	Habitat Restored	Acres	Emissions Reductions $MTCO_2eq$/ Acre/Yr.	Carbon Sequestration $MTCO_2eq$/ Acre/Yr.	Projected $MTCO_2eq$ over 80-year Period
Mississippi Alluvial Valley	Bottomland Hardwood Forest	50,000	2.213	5.831	32,174,551
Northern Great Plains	Grassland	35,000	0.999	1.485	4,357,215
Chesapeake Bay	Riparian Forest	2,000	0.768	4.528	856,359
East Texas & Oklahoma	Bottomland Hardwood Forest	3,000	2.213	5.804	1,930,473
Great Lakes	Bottomland Hardwood and Riparian Forest	5,000	0.774	4.584	2,143,303
Puget Sound	Wetland Forest	5,000	0.994	9.534	4,213,424
Totals		**100,000**			**45,675,325**

* This portfolio represents projects for which DU has immediate capacity to begin comprehensive habitat restoration, once funding is in place.

COST FACTORS

Terrestrial carbon sequestration is much more cost effective than other methods. Reducing emissions via geologic sequestration or smokestack removal costs \$20–\$150 per $MTCO_2eq$. In comparison, restoring land that captures carbon and reduces emissions associated with agricultural operations costs \$3–6 per $MTCO_2eq$. Even after taking into account the likelihood that most carbon benefits are not realized until the 10th to 40th year period of tree growth, forest projects are estimated to cost \$10 per $MTCO_2eq$.

These cost estimates do not include a number of factors unique to terrestrial carbon sequestration projects. On the positive side are the payoffs from other eco-asset benefits associated with land restoration, such as water quality enhancement and habitat protection. On the expense side, there are additional costs to consider in education and outreach as well as assisting landowners and farmers to make a major shift in "standard" practices.

On most U.S. agricultural land, a major management shift is required in the change to a no-till system, bringing along substantial risk. Methods to help farmers mitigate these risks must be addressed, and training should be provided on how to successfully change management systems. In addition, it will be important to address the interests of farmers and others who already utilize carbon sequestering management practices and whose soil is already richer in carbon. They may not have the same opportunities to sequester additional carbon, and it is very important that they should not be punished or passed by on rewards because of their previous good stewardship. In the end, financial incentives, such as carbon easement payment could increase farmer's profit and reward them for good management practices.

Financial incentives such as conservation easements are important. It can be difficult to get a farmer to commit to these projects without upfront payment. This can be a catch-22 inasmuch as land must be committed for restoration before projects can be presented to industry to fund. Thus, carbon offset aggregators may spend much time explaining the carbon market and getting farmers motivated to enroll their land, but are not able to bring money until after a transaction takes place.

Finally, when determining the cost of carbon offset projects to farmers and carbon aggregators, expenses on carbon measurement, monitoring, and verification have to be taken into account. This will increase the cost per metric ton needed to attract landowners to enroll land in a carbon easement in order to ensure the adequate implementation of baseline and monitoring procedures.

CONCLUSION

If structured properly, terrestrial carbon sequestration projects can have a significant and immediate impact on the carbon market, thereby providing vital time needed to develop new low-emission fuel sources with which to meet the world's growing energy needs. Land restoration and land management often generate multiple enhancements to the environment beyond the benefits of carbon sequestration—including such enhancements as water quality, wildlife habitat, mitigation banking, and forest banking. Partnerships between energy companies and conservation organizations will help define this market and enable industry to demonstrate a commitment to environmental stewardship to its consumer base, regulators, U.S. Congress, international treaties, and the general public while providing critical revenue to farmers and rural communities struggling with a depressed agricultural economy.

References

Birdsey, R. A. 1996. "Regional Estimates of Timber Volume and Forest Carbon for Fully Stocked Timberland, Average Management after Cropland or Pasture Reversion to Forest." Appendix 3 in *Forests and Global Change,* vol. 2, *Forest Management Opportunities for Mitigating Carbon Emissions.* R. N. Sampson and D. Hair (eds.). Washington, D.C.: American Forests.

Euliss, N., R. Gleason, A. Olness, R. McDougal, H. Murkin, R. Robarts, R. Bourbonneire, B. Warnes. 2002. *Prairie Wetlands of the United States Important for Carbon Storage: Implications for Future Research and Management.* Washington, D.C.: Agricultural Research Service, USDA.

FAO. Land cover classification system: Classification concepts and user manual (with CD-ROM)). Rome, 2000. 179 pp. ISBN 92-5-104216-0.

Follett, R. F., J. M. Kimble, and R. Lal. 2001. "The Potential of U.S. Grazing Lands to Sequester Soil Carbon." In R. F. Follett, J.M. Kimble, and R. Lal. (eds.). *The Potential of U.S. Grazing Lands to Sequester Carbon and Mitigate the Greenhouse Effect.* Washington, D.C.: Lewis Publishers.

Hodgetts, R. 2000. U.S. Department of Agriculture and Forestry Agriculture Statistics.

Washington, D.C.: Intergovernmental Panel on Climate Change (IPCC). 2001. Third Assessment Report—Climate Change.

International Union for Conservation of Nature and Natural Resources (IUCN). 2002. Climate Forests and People Information Desk. http://www.iucn.org/themes/carbon/index.htm

Lal, R., J. M. Kimble, R. F. Follett, C. V. Cole. 1998. *The Potential of U.S. Cropland to Sequester Carbonand Mitigate the Greenhouse Effect.* Chelsea, MI: Ann Arbor Press.

Lewandrowski, J., M Peters, C. Jones, R. House, M. Sperow, M. Eve, and K. Paustian. March 2004. Economics of Sequestering Carbon in the U.S. Agricultural Sector. *Technical Bulletin* No. 1909 (TB1909), 68 pp.

Lower Mississippi Valley Joint Venture. 2002 (unpublished). *Conservation-based Incentives for Terrestrial Carbon Sequestration in the Lower Mississippi Valley: Issues and Options for Integrating Government-financed and Private Sector-financed Programs.*

National Aeronautics Space Association (NASA). March 2004. Land Cover Changes Affect U.S. Summer Climate. Goddard Space Flight Center.

Sperow, M., M. Eve, and K. Paustian. 2003. Potential Soil C Sequestration on U.S. Agricultural Soils. *Climatic Change* 57 (3), April: 319–39.

U.S. Department of Energy National Energy Technology Laboratory (April 2004). *Carbon Sequestration: Technology Roadmap and Program Plan.* Washington, D.C.: National Energy Technology Laboratory.

U.S. Department of Energy. April 2004. *Carbon Sequestration—Technology Roadmap and Program Plan.* Washington, D.C.: Office of Fossil Energy.

U.S. Environmental Protection Agency. April 2003. *Inventory of U.S. Greenhouse Gas Emissions and Sinks:* 1990–2001. Washington, D.C.

U.S. Department of Agriculture. 2004. *U.S. Agriculture and Forestry Greenhouse Gas Inventory: 1990–2001.* Global Change Program Office, Office of the Chief Economist. *Technical Bulletin No. 1907* (TB1907), 163 pp.

Wear, D. and J. Greis. 2001. *Southern Forest Resource Assessment Report.* USDA Forest Service Southern Research Station.

West, T. O., and G. Marland. 2002. *A Synthesis of Carbon Sequestration, Carbon Emissions, and Net Carbon Flux in Agriculture: Comparing Tillage Practices in the United States.* Oak Ridge, TN: Oak Ridge National Laboratory.

---

[1] Fertilizer produces nitrous oxide, a very powerful greenhouse gas, approximately 296 times more potent than carbon dioxide (USDA).

Dick Kempka is *director,* Energy and Technology Partnerships for Ducks Unlimited, Inc. and Dawn Browne is Geographic Information Systems Manager. Mr. Kempka can be reached at T: 901-758-3795 and E: dkempka@ducks.org.

CHAPTER 15

Information Technology: Enabling and Accelerating Environmental Markets

By William G. Russell

Environmental Markets are an essential economic component of society's environmental sustainability. They will allow many environmental and social risks, including their full costs to be better attributed to the companies, products and services that cause them and benefits to those that mitigate and resolve them. As environmental markets mature, traditional markets and market economic values will gradually and continually adapt to reflect society's integrated environmental social and economic needs. These values are further adjusted to reflect our capacity to develop and implement innovative technologies and sustainable solution alternatives.

Our capacity to develop and implement Environmental Markets can be accelerated by observing and incorporating the experiences, lessons and best practices derived from technology-based markets. Emerging technology trends provides confidence in our ability to design and implement complex business management systems and associated environmental markets. Sustainability-driven systems and markets are both technically and economically achievable.

This chapter introduces some of the lessons derived from technology-based markets. It also discusses selected technology trends that will enable complex sustainability-driven systems such as the environmental markets. Very importantly, this chapter has not addressed cultural changes, both within corporations and in the marketplace, that are necessary for the successful adoption of sustainability-oriented technologies and full development of the environmental markets. Critical cultural change success factors include: leadership, systems-based problem-resolving archetypes, collaboration, copyright protection and intellectual property valuation, privacy protection and confidential information. These success factors deserve a dedicated chapter.

ACCELERATING ENVIRONMENTAL MARKET DEVELOPMENT: ANALOGIES FROM THE INFORMATION TECHNOLOGY MARKET

In 1991, Geoffrey Moore published an insightful book, *Crossing the Chasm.* Forecast to sell 5,000 copies, it has sold hundreds of thousands of copies and subsequent sequel publications. Today and for the foreseeable future, it will be considered required reading for every high-tech executive, both in and outside the information technology (IT) sectors. The book's appeal was that "it puts a vocabulary to a market development problem that has given untold grief to any number of high-tech enterprises." This previously non-articulated problem was eloquently stated as *Crossing the Chasm.*

Technology Adoption Lifecycle

The environmental sustainability approach, enabling technologies, and associated environmental markets are only in the beginning stages of their adoption lifecycle. At this time, they are being defined and promoted by just a few leaders within science, business, government and other stakeholder groups. Figure 1 presents The Chasm Group's Technology Adoption Lifecycle.

Figure 1

Technology Adoption Lifecycle

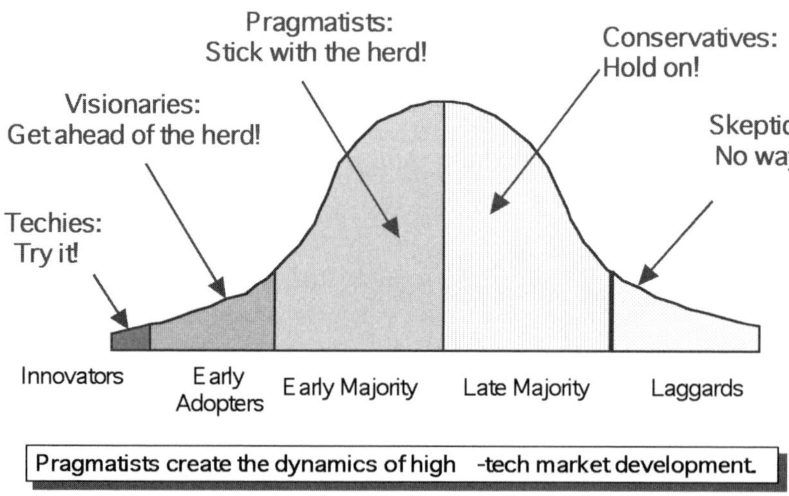

Source: The Chasm Group

The question is not if environmental markets will develop, but which ones and how soon. Smart businesses will know their sustainability impacts and foster corporate cultures that systematically search for innovations, reward strategic risk taking, and motivate rapid change. As in nature, the laggards will diminish and in time become extinct.

Market Development Models

Using the Sustainability Adoption Lifecycles, how might environmental-market innovators better define their markets and accelerate the "herd" to adopt? Some insights can be gained by examining the Chasm Group's high-tech Market Development model (Figure 2).

Figure 2

Market Development Model

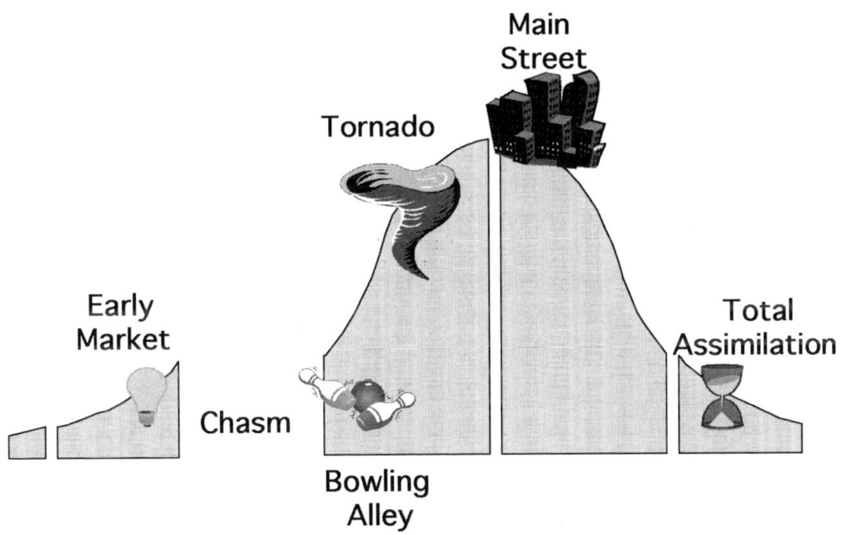

Source: The Chasm Group

The model has segmented the adoption lifecycle into six zones. Business strategies, sustainability applications, and environmental markets mature as their respective integrated marketplaces move through these stages.

1. **The Early Market**, a time of great excitement when customers are enthusiasts and visionaries looking to be first to get on board with the new paradigm.
2. **The Chasm**, a time of great despair, when the early-market's interest wanes but the mainstream market is still not comfortable with the immaturity of the solutions available.
3. **The Bowling Alley**, a period of niche-based adoption in advance of the general marketplace, driven by compelling customer needs and the willingness of vendors to craft niche-specific whole products.
4. **The Tornado**, a period of mass-market adoption, when the general marketplace switches over to the new infrastructure paradigm.
5. **Main Street**, a period of aftermarket development, when the base infrastructure has been deployed and the goal now is to flesh out its potential.
6. **End of Life**, which can come all too soon in rapidly changing high-tech markets where entirely new paradigms come to market and supplant the leaders who themselves had only just arrived. End of Life takes on a slightly more critical connotation to those of us working on Sustainability. The extinction of laggard Corporations is a positive outcome. The extinction of human life (as we know it) is an alternate potential outcome.

IT TRENDS ENABLING ENVIRONMENTAL MARKETS

Understanding trends in information technology platforms, data storage and knowledge management can provide perspectives for evaluating current products and additional insights as to how they could influence the design and adoption lifecycle of sustainability-oriented applications such as environmental markets.

Information Technology Platforms

An essential component for designing and developing successful environmental markets is the development and adoption of new information technology applications that enable sustainability knowledge management and performance improvement. Technology Platforms determine the infrastructure and communications standards necessary to enable collaborative knowledge development, deploy complex business management and reporting systems and implement the performance improvement activities required to achieve sustainability. IT platforms have rapidly migrated from their initial PC-based decentralized applications to enterprise level client server networks, through today's almost seamless use of extranets and the Internet. These technology advancements demonstrate the importance of managing technology adoption lifecycles and the accompanying cultural adjustments. Such management determines whether users will embrace or resist the adoption of new technologies and their intended performance and productivity improvements.

Shared Servers and Services

Shared Servers and Services is a critical technology trend that allows companies to consolidate and leverage hardware, software applications and databases investments across multiple customers and supply chains. Initial applications have dramatically enhanced productivity, and reduced the hardware and software costs. Shared servers and services have also proven to be disruptive to an organization as repetitive tasks are automated and outsourced, and individuals become more concerned about information security, confidentiality and privacy.

Knowledge Management Technologies and Portal Services

Knowledge management technologies are intended to solve the problem of having too much data and information available from multiple sources to be humanly managed. They automatically transform information-overloaded systems to ones that deliver the right information to the right user at the right time. Information can be filtered based upon a number of criteria such as a user's industry context, company-specific needs, role in the enterprise and the given task at hand.

Knowledge management technologies were quickly assembled into more user-friendly "whole" products- or knowledge-oriented ASP portals. These portals took many of the features of knowledge and content management, and combined them with powerful search engines and intuitive, customizable Web interfaces. Portals that have emerged include the following varieties:

- **Collaborative Portals**—Applications that focus on communication features such as team rooms, project management tools, discussions, chat rooms, and e-mail.

- **Expertise Portals**—Provide informed advice, decisions, or recommended solutions. Serves as librarian, advisor, instructor, and general assistant by extending the bounds of what can be automated to include tasks that require reasoning, symbolic processing and problem solving.
- **Knowledge Portals**—Combine the components of the above market segments but with a concentration on:
  - ➢ Retrieving core information from corporate IT systems or external sources.
  - ➢ Managing it according to preferences, roles, and specific tasks of individual users.
  - ➢ Facilitating communication and collaboration between those who can supply information and those who need it.
- **Enterprise Information Portals**—These portals forsook their knowledge management ancestry in favor of providing centralized browser-based access to databases, applications and other enterprise systems.
- **Solution Portals**—As enterprise portals migrate from client server platforms to the Internet and incorporate the more efficient shared server and services business model, complete solution portals will be assembled.

While many organizations are struggling with how to classify knowledge and implement knowledge management programs with as little disruption as possible, those who are proactive look to knowledge management as a fundamental element of competitive advantage.

Technology Applications and Portals Landscape

Traditional management systems were not designed for a balanced view of financial, environmental and social metrics. They were developed to measure performance data for quality-, risk-, and cost control. Next-generation tools for sustainability go beyond compliance and risk management. New technologies promise clearer views of complex issues with more integrated functionality. They are influencing new thinking about the causal relationships of issues and are forcing a shift in decision-making and management behaviors in favor of reduced burden-shifting and greater accountability. Figure 3 provides a summary of the current application and portal landscape.

Figure 3

Application and Portal Landscape

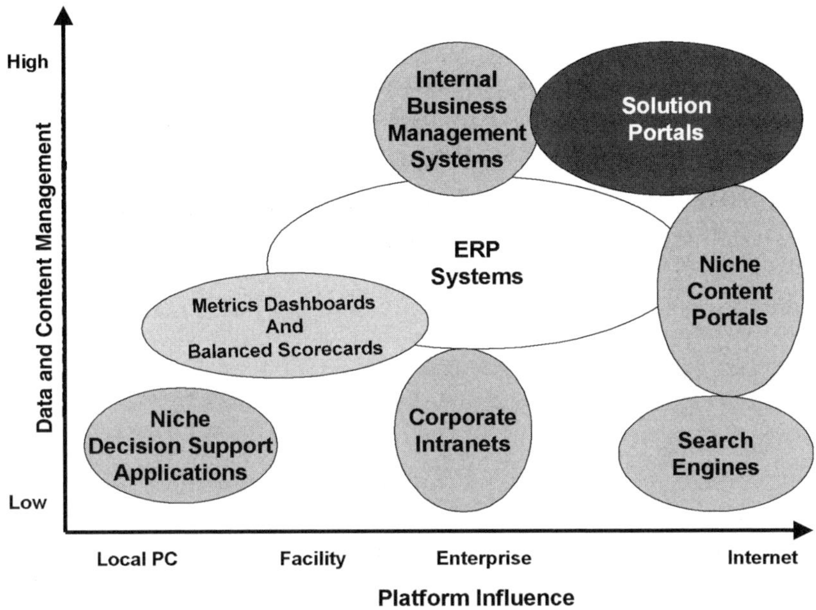

The technology trends have manifested themselves into a range of technology product categories. These products incorporate technology developments related to the collection and analysis of data referred to in the technology world as structured content, and the management of unstructured content such as documents or video. The major functions of business management systems today are data management and business process automation or work flows. The major functions of application service providers are aggregating unstructured content, and adding new value to this content through the use of knowledge management and collaboration technologies. While the future will certainly see the integration of these product categories, they provide a strong basis for evaluating the needs and opportunities of sustainability business systems and environmental markets.

Whatever solution a company chooses, the most important thing to keep in mind is the flexibility of the portal system. We are already anticipating further development trends toward full service or "whole" solution portals. If at any time it seems that the portal will force a business to change to meet its requirements, then it's time to move on. A portal solution works only when it can easily coexist with corporate information and culture, accepting and integrating content wherever it comes from and however it is created.

BUSINESS MANAGEMENT APPLICATIONS AND SYSTEMS: ENABLING DATA COLLECTION, ANALYSIS, WORKFLOW AUTOMATION AND REPORTING

Effectively managing Sustainability and strategically leveraging environmental markets requires a very comprehensive enterprise-level tracking system for everything that impacts long-term ability to operate with a net-positive result in all dimensions. Many companies are realizing that, like with Sarbanes-Oxley compliance, they can theoretically attempt to implement Sustainability using a manual paper-based system. However, in practical terms, the only way to achieve their objectives is to use powerful integrated analytical software, database and knowledge management tools. No manual system could ever hope to equal the degree of integration and automation currently possible with well-designed technology solutions.

The current products advancing sustainability solutions today embrace more simplistic market definitions and solution implementations, and were not developed to meet sustainability needs. Current products are developed to address issues such as automating six sigma and balanced scorecards; supply chain management; and regulatory compliance. In energy management and GHG emissions inventory programs, these products are advancements when compared to today's business as usual management systems. Much opportunity remains for those products that successfully integrate existing technology functionality and business management functions within a more comprehensive "whole" product solution.

Standards for Management Systems and Environmental Markets

Sustainability and environmental markets are still only recently adopted concepts in the business world, so there currently exists a shortage of tools available to adequately exploit their benefits at an enterprise level. Having market standards and uniform analysis methodologies inform a system's design and greatly improves the utility of the information being processed. The following examples illustrate what is currently available and the expected direction of near-term product enhancements.

The Greenhouse Gas Protocol Initiative (www.ghgprotocol.org) is a multi-stakeholder partnership convened by the World Resources Institute (WRI) and the World Business Council for Sustainable Development (WBCSD). The Initiative's mission is to develop internationally accepted greenhouse gas (GHG) accounting and reporting standards for business and to promote their broad adoption. The GHG Protocol Initiative comprises two separate but linked standards:

- **GHG Protocol Corporate Accounting and Reporting Standard** which provides a step-by-step guide for companies to use in quantifying and reporting their GHG emissions, and
- **GHG Protocol Project Quantification Standard**, a guide for quantifying reductions from GHG mitigation projects.

These standards have been designed to be program or policy neutral. However, many existing GHG programs use it for their own accounting and reporting requirements. To complement these standards, a number of cross-sector and sector-specific calculation tools are available on the GHG Protocol Initiative website. These tools provide step-by-step guidance and PC-based electronic worksheets to help users calculate GHG emissions from specific sources or industries. The tools are consistent with those proposed by the Intergovernmental Panel on Climate Change (IPCC) for compilation of emissions at the national level (IPCC, 1996).

SANGEA™ Emissions Estimation Application for the Oil and Gas Industry The SANGEA™ system is the result of an effort by ChevronTexaco, aided by the American Petroleum Institute (API), to develop more useful and accurate sector specific accounting methods for the worldwide oil and gas industry. This tool and associated guidance documents for calculating greenhouse gases emissions from the production of oil and gas will represent a substantial addition to the tools previously available to companies conducting GHG inventories. The worksheets are intended to provide producers with calculation methods that follow the American Petroleum Institute's Compendium of Greenhouse Gas Emissions Estimation Methodologies for the Oil and Gas Industry.

ChevronTexaco developed the software to provide an auditable, Excel-based tool for estimating greenhouse gas emissions and energy utilization, and linked the software to a specific relational database to facilitate data management at the corporate level. The inventory coordinator at each facility configures a spreadsheet, enters monthly data and sends quarterly reports to the database. The SANGEA™ system has improved the efficiency and accuracy of the company's greenhouse gas inventory.

ChevronTexaco's onetime investment in developing the SANGEA™ system is expected to reduce the long-term expense of maintaining a legacy system or hiring independent consultants. The SANGEA™ system enables the company to standardize on methodologies for quantification of emissions, while the WRI/WBCSD Protocol provides standardized boundaries for inventory accounting. Replacing a diverse and confusing set of accounting and reporting templates throughout the company has yielded efficiency and accuracy gains that now allow the company to more accurately manage GHG emissions and institute specific emissions improvements. Standardization has driven these improvements. ChevronTexaco's experience with the SANGEA™ system reveals the value of intelligent and consistent GHG accounting and management. See www.ghgprotocol.org for instructions on how to obtain a copy of the software.

International Performance Measurement & Verification Protocol (IPMVP) (www.ipmvp.org) provides standard measurement and verification (M&V) terminology, and defines four M&V options to quantify energy and water savings. It is a savings-verification tool with principles that are applicable to commercial and industrial energy and water efficiency projects. These standards were initiated by the U.S. Department of Energy (DOE) and facilitated a market transformation initiative to help secure low-interest loans from financial institutions for energy efficiency investments. Seven years and three editions later, these standards are now maintained through a non-profit organization, IPMV, Inc. They have become standard in almost all energy efficiency projects where payment to the contractors is based on the energy savings.

Global Reporting Initiative (GRI) (www.globalreporting.org/) GRI as a global facilitator of sustainability reporting standards, GRI also understands that software based reporting tools offer an important contribution to sustainability reporting. GRI has partnered with a number of organizations to create a series of tools geared toward bridging the existing "delivery" gap between GRI report writers and report readers. GRI's reporting portal is striving to create the following:

- **Central repository:** This will be a relative database containing all information reported, it will be accessible through the internet and will be populated via automated data input as reporters compile their data.
- **Reporting wizard:** Mainly envisioned to be a report writing tool with basic functionality. Data entered here will be uploaded to the central repository.

- **Centre for information exchange:** This will house all GRI reporting tools and act as a collaborative platform.

ENVIRONMENTAL MARKET IMPLICATIONS AND OBSERVATIONS: WHAT ZONE ARE YOUR ENVIRONMENTAL MARKET DEVELOPMENT EFFORTS IN?

To complete this chapter's lifecycle, it is helpful to check in with the six market development zones applied by Geoffrey Moore and The Chasm Group.

- **No matter where you think you are, Early Market excitement is still appropriate.** *The Early Market,* a time of great excitement when customers are enthusiasts and visionaries looking to be first to get on board with the new paradigm.

  Sustainability and environmental market innovations are all effectively in Early Market positions. Early-stage environmental markets will see some associated benefits as Sustainability business management and portal products are adopted as well as from aligning and leveraging developments within traditional capital markets. The paradigm shift toward new Sustainability-oriented market designs and, more importantly, market values has begun. Traditional capital market tangible value measurement methods are expanding to incorporate more intangible values.

- **Look at the Chasm, but then make the leap across.** *The Chasm,* a time of great despair, when the early-market's interest wanes but the mainstream market is still not comfortable with the immaturity of the solutions available.

  Energy trading markets have been harmed by human failures, early market inefficiencies, and a lack of clear market objectives and standards. Greenhouse Gas Emission Reduction Markets have been following the roller coaster ride associated with complex global treaty negotiations, competing technology interests and a lack of clear market objectives and standards.

- **Bowling a strike requires throwing the ball, hitting the headpin and generating action.** *The Bowling Alley,* a period of niche-based adoption in advance of the general marketplace, driven by compelling customer needs and the willingness of vendors to craft niche-specific whole products.

Socially-responsible investing has crossed the chasm, with companies and index-specific financial products moving through sector-specific adaptations. Market leaders must enhance their current technology infrastructures and product offerings to prepare for their long awaited Tornado growth opportunities. U.S. Cap and Trade markets for NOx and SO_2 are ripe to be extended to new industry sectors, and new credit rights are being considered to cover other environmental liabilities. GHG credit trading systems, designed and implemented by individual countries, states or even emissions-source types, are all likely examples of environmental market bowling pins.

- **It may seem calm, but prepare for the Tornado.** *The Tornado* is a period of mass-market adoption, when the general marketplace switches over to the new infrastructure paradigm.

  It is fair to say that neither the Sustainability management systems nor the environmental markets are quite at this stage. However, some niches are better than others in being positioned to lead the "herd" and it would wise to keep an eye out for them.

- **Environmental Markets will thrive from Wall Street to Main Street.** *Main Street,* a period of aftermarket development, when the base infrastructure has been deployed and the goal now is to flesh out its potential.

- **Cradle-to-cradle alternative to end of life.** *End of Life,* is a well-understood outcome in rapidly changing high-tech markets.

  Paper-based and human operated trading systems will migrate to technology-based trading platforms. Traditional commodity markets must adapt as they may experience significant changes or be replaced by new "cradle" material- and energy-integrated environmental resource markets that are more aligned with ecosystem services and social values.

William G. Russell is *CEO* of SKN Worldwide. He can be reached at T: (201) 592-0055 and E: wgrussell@sknworldwide.com.

CHAPTER 16

Green Trading Markets: Where Are We Now?

By Peter C. Fusaro and Marion Yuen

In this collection of essays, we find discussions of green trading developments primarily in the United States and include examples from Australia, Italy, and Eastern Europe. Throughout the chapters, we see the beginnings of a new phase, a second wave of Green Trading. So, where are we now in 2005? The second wave of Green Trading is taking off with the implementation of the Kyoto Protocol and the European Union Emissions Trading Scheme (EU ETS). Projects and trades have begun, and much institutional money has flowed into project-based reductions. In the U.S., there is the interesting marriage of CO_2 injection for enhanced oil recovery and carbon sequestration. And very importantly, in December 2004, there was a breakthrough on Wall Street in the Fitch Rating's Special Report on Emission Trading.

Willingly or unwillingly, with ease or with angst, the old must give way to the new. Currently, our economic and social structures are simultaneously responding to a fundamental shift in fuels and the effects of global climate change. In this transition to a new, carbon-constrained global economy, we are challenged to depart from the known and familiar, and to re-examine our values, standards, and practices. Along the way, economic values will be distributed differently and new kinds of competition will emerge as environmental factors become key in business decision-making and carbon promises to be the new gold. We need money and intellectual capital to remediate environmental degradation and create an environmentally benign world. In this changeover, Green Trading has created commodities that provide economic incentives for growth as new paradigms evolve.

Proposed in the United States by the Environmental Defense (a U.S. environmental organization) in the early 1990s, environmental financial trading began to enter into its second phase within less than 10 years. In 2002, we coined the term, GreenTrading™, to promote the Triple Convergence of trading in credits associated with greenhouse gas emissions reduction, renewable energy, and energy efficiency (negawatts) through the use of the financial markets. In this Triple Convergence, environmental financial risk is treated as a mainstream corporate financial issue. The intent of GreenTrading™ is to capture both the problem and the solution, with financial trading providing a means to ameliorate pollution. Building on the successful U.S. sulfur dioxide emissions trading program, the long-term impact of such trading would be to reduce pollution in a cost-effective manner and accelerate the introduction of more environmentally benign technologies. It would decrease economic disruption to the capital-intensive energy industry and other industrial sources of pollution. At the same time, it would create new financial markets where "trading pollution," as it is sometimes mistakenly called, would actually create concrete and measurable emissions reductions for American business. This business model could be exported throughout the world, and it has happened to some degree. Indeed, the generic term, Green Trading, has begun to be used by leading practitioners around the world.

Already in the second wave of Green Trading, we see an expansion in coverage and types of players coming into the markets. For instance, the EU ETS' allowance system covers thousands of facilities in multiple industries with varying degrees of control costs, providing the basis for a good market. Initial transactions have occurred between Shell and the Dutch utility, Nuon, and between Shell and Barclays, representing the participation of energy and financial firms. Also, there was a very large trade of 10 million tonnes of Certified Emission Reductions (CERs) between Nuon and AgCert on Brazilian CDM credits in the 2005–2007 period. In the U.S., the Chicago Climate Exchange (CCX, launched in September 2003) now has over 70 members trading GHG on a voluntary basis (see Chapter 5) and there are many other companies that have started to look seriously at self-imposed GHG caps. CCX has expanded into Europe with the European Climate Exchange and linking with the electronic platform of International Petroleum Exchange, the well-established energy futures exchange. In the corporate environment, we are starting to see the risk manager handle the GHG issue, and carbon finance playing a role in asset decisions. So, we are witnessing the incorporation of these commodities into the marketplace.

BREAKTHROUGH ON WALL STREET

Environmental financial risk is rising as an issue in corporate America and throughout the world. As stated in Chapter 9, "traditional commodity forecasting methodologies are producing conclusions that fail to address the characteristics of the GHG market and fail to effectively guide corporate policy development."

The December 2004 Fitch Rating's Special Report on Emission Trading marked the beginning of a new view on Wall Street. The report draws the following conclusion on the corporate ratings implications of emissions trading: "Fitch anticipates more stringent pollution-control requirements leading to increased operating and capital costs. A well-structured emission-trading program can assist companies in managing and reducing capital expenditures for compliance with environmental regulations."

Fitch views that "the commencement of carbon (CO_2) trading in Europe and the recent run-up in prices in the United States for sulfur dioxide (SO_2) emission credits have contributed greatly to renewed attention to value and effectiveness of the use of emission credits. ... Fitch also recognizes that a trading program that does not foster the overall reduction in pollutants would ultimately fail. Historically, the use of allowances alone has not been the most cost-effective way of dealing with environmental problems, and this is likely to remain the case. The purchase of emission credits is one part of a solution that includes installation of pollution-control devices, fuel-switching, conservation or demand management, improved efficiency." Further, Fitch "believes that the United States will eventually have a federal law limiting the emissions of CO_2 and possibly other greenhouse gases (GHGs)." This is the new consensus that is rising in corporate America.

The Fitch report, followed shortly by the implementation of the European Union's Emissions Trading Scheme and the Kyoto Protocol, marks an acceleration of the trend towards putting climate change risk on the balance sheets of corporations throughout the world. It is a harbinger of the shifts in thinking, that must happen in the ratings agencies and the rest of the capital markets.

CORPORATE LEADERSHIP

Not only has environmental risk become a financial issue, it is also a corporate fiduciary responsibility and an increasing shareholder concern. As companies analyze their risk, they have begun to realize that there is a global issue here and that they need to do something.

The energy and agriculture industries—the world's leading air polluters—are the logical and likely leaders in providing environmental solutions because it is good for their business. Together, these industries constitute the world's largest businesses, and very significantly, they also have the most deeply liquid commodity markets. In particular, the energy industry has the financial strength, intellectual capital, and global presence to provide these solutions (see Chapter 10).

Since the private sector has a vested commercial interest in emissions reduction, it will take the lead in the development of emissions trading markets. One of the drivers behind the GHG market is the movement of institutional shareholder actions forcing corporations to acknowledge the environmental risk on their books. Shareholder resolutions led by pension funds have adopted strategies similar to that taken for tobacco regulation in the U.S. So far, these have produced an impact on the leading U.S. oil and gas companies: in March 2005, six of these corporations agreed to disclose their potential financial exposure from climate change and embark on a range of actions to reduce exposure to climate risk. On the global level, institutional investors (with $20 trillion assets) in the Carbon Disclosure Project have continued to put pressure on companies for disclosure of information on their environmental performance.

As the environmental financial markets develop, the principals for market making will be the investment banks, multinational oil and gas companies and agribusiness. The latter two sectors will be important suppliers of liquidity since both are already active in commodity trading. In coming years, corporations will have global environmental portfolios managed with profit-and-loss targets. In recognition of the leading role that the private sector must play in the post-Kyoto era, the Pew Center for Global Climate Change has specifically included energy-intensive industries among the stakeholders in its research discussions to explore a next generation framework.

ENABLING REGULATORY ENVIRONMENT

The United States has the world's most mature environmental financial markets in SO_2 and NOx emissions allowances and most accumulated trading experience. Further, it was the U.S. delegation that introduced emissions trading into the international climate change process.

Cap-and-trade markets-based tools have been proven to work in reducing SO_2 and NOx pollutants. What we need now are structurally sound market-based tools that cover Green Trading commodities—CO_2 and other GHG gas (methane, nitrous oxides, hydrofluorocarbons, perfluorocarbons, and sulfur hexafluoride) emissions reduction as well as renewable energy and the financial value of energy efficiency efforts. The end result would be fungible commodities that could trade anywhere in the world. Today, we are a long way from that goal and in the process of creating the new markets, the public and the environmental community will be watching whether the market-based mechanisms are actually effective in reduction pollution. At the same time, companies will need incentives to take action now and the confidence that they will be rewarded, or at least not penalized, for early actions as regulations take time to evolve.

The SO_2 markets have a 35-year regime of reductions and increasingly stringent standards. To effectively reduce CO_2 and other GHG gases, there needs to be a 100-year program that engages the entire world. Longer-term targets, lasting into decades, provide corporations with the necessary periods of certainty for investment planning and infrastructure building. Further, since CO_2 disperses into the atmosphere on a global scale, it is necessary to broaden beyond the Kyoto Protocol to involve the developing world and the United States which contributes 25 percent of global GHG emissions.

There is no quick technological fix as long as the world is addicted to fossil fuels. That habit is not going to change quickly as it typically takes years to implement the alternatives required to put meaningful dents into CO_2 emissions. To treat CO_2 emissions seriously, we need a regime that will aggressively reduce global carbon intensity from both stationary and mobile sources, accelerate technology transfer, and increase energy efficiency. The irony is that the technology exists today to get the job done and not pie in the sky proposals about the hydrogen economy of tomorrow. Instead, we have highly efficient integrated gas combined cycle (IGCC) technology for coal gasification as well as other fossil fuels such as petroleum coke, wood chips and municipal solid waste now. We have affordable hybrid vehicles that reduce both tailpipe emissions and fuel economy now. We have many energy efficiency devices that reduce building loads from both commercial and residential buildings. And certainly, there is the controversial nuclear option. We do not need to keep issuing studies that predict ecological catastrophes. Rather, we need to take action now, that will create economic development and jobs around the world as an environmental financial industry develops and other industries rise in response.

So, the solutions exist. However, for many of them to become commercially viable in the near term, we need the government to establish reasonable policies that bring a financial value to emissions reductions so that these solutions work for both business and the environment. The point is that both the SO_2 and NOx programs are mandated and have financial penalties for noncompliance. These real financial consequences have allowed technologies such as scrubbers and low-NOx burners to take hold, and they can also spur the accelerated adoption of IGCC technology. Voluntary CO_2 programs may be useful in practicing for future global trading of environmental financial credits but hard limits will be needed to create a real market driver for change.

Emissions trading has proven to be an effective mechanism to accomplish much of the environmental goals in economical ways. Further, the ability to monitor and certify verifiable reductions is already in place through both third-party certification companies as well as the use of geopositioning satellites and remote-sensing devices. Financial markets for the environment *work*.

BUILDING ON SUCCESSFUL EXPERIENCE

The evolving regulatory landscape is still an open issue. On the horizon, we expect more states to consider and enact renewable portfolio standards (RPS) and GHG reduction systems. Presently, 28 states are acting on GHG and 19 states have RPS. In the U.S. SO_2 program, we saw something that we might see for GHG and for renewables. Because so many states started to put together their own regulations that companies operating in a multistate environments finally told the federal government they wanted some consistency in the regulations. That demand resulted in the 1990 Clean Air Act amendments which authorized the first successful emissions trading program for SO_2.

The GHG markets can be expected to take off and follow a similar rate of acceleration as the SO_2 and NOx markets experienced in the U.S. This is because there is more at stake and because the European market can draw on the U.S. experience. In addition, the GHG markets can also draw on the experience and the talented pool of people that are available in the financial community and the energy trading community. The U.S. is still well positioned to lead on environmental financial market development with its entrepreneurial culture, risk capital, and knowledge base in trading. Moreover, U.S. multinational companies active in Europe are now in the vice of dual environmental standards, i.e., one for Europe and one for the U.S. This is an untenable position for corporate America.

CONSISTENT AND COMMON ECONOMIC TOOLS

An emissions trading program is primarily valuable as it puts a market price on the cost of emissions, thus allowing a company to make an informed choice among compliance options. However, it should be noted that a dysfunctional trading system would result in market prices that could lead to economically suboptimal decision-making. For an emission trading market to be efficient, target levels of emissions must be assigned in a consistent and coordinated fashion. With the U.S. SO_2 program, there is one overarching regulatory body in the U.S. Environmental Protection Agency (EPA) which allocates the allowances and monitors the compliance process. However, with the carbon markets, a number of countries would be setting their own procedures with varying degrees of rigor.

Consistency and compatibility of schemes will also be central. Changes in trading rules are disruptive to any market and could lead to sharp changes in the value of allowances. Inconsistent application of schemes could leave companies either overinvesting in certain technologies relative to future needs, or investing today in remedial technologies based on a cost-benefit analysis using output requirements that might increase or include additional pollutants over time. Further, for emissions reductions programs in which permissible emissions levels are reduced over time, banking of credits/allowances is important so that those credits/allowances generated in the early years could be used for compliance in later years. Banking provisions give generators the incentives to reduce emissions more rapidly than required and allow them greater flexibility in capital expenditures.

Some financial institutions such as Morgan Stanley have been trading SO_2 and NOx emissions reduction allowances as these are a natural compliment to investment banks' sizable presence in the electric and natural gas markets. In addition, a small number of brokers have established niches in the Green Trading markets, including Amerex, Cantor Fitzgerald, EcoSecurities, Evolution Markets, GT Energy, United Power, and Natsource.

It is anticipated that the U.S. regulated futures exchanges will participate in trading and clearing environmental contracts. Already, the New York Mercantile Exchange has indicated the intention to trade and clear SO_2 and NOx futures contracts. The advantage of clearing trades through an exchange is that it increases price transparency as prices are posted on a real-time basis. Additionally, it reduces counterparty risk as the exchange steps in between the market participants.

A number of voluntary, bilateral sales of GHG reduction credits have taken place, principally under auspices of major international and domestic corporations. Although several schemes are already in place or under development, there is no global standard for verification of these transactions. Neither is there systematic, comprehensive, and consistent recording of these reductions among registries. Implementation and use of economic tools that operate across and among registries would encourage financial and physical transactions, and establish a more narrow value for GHG reduction credits.

Current U.S. policy promotes voluntary GHG reductions, usually carbon dioxide emissions reductions. Hence, we have the voluntary Chicago Climate Exchange. Moreover, federal standards for mandatory reduction in GHG emissions are not being considered by the current Administration. Nevertheless, an efficient, powerful federal registry would be very helpful in ensuring development of active bilateral and private exchange trades in carbon dioxide reductions as well as allow entities to bank their reduction credits for future use or private voluntary sales, or register their current carbon footprint in order to take advantage of any early mandates aimed at GHG reductions. However, the federal government is not actively considering such a registry.

In response, a number of U.S. states and regions are now developing their own GHG registries and renewable energy standards, but these efforts are not uniform with regard to many metrics, rules, and protocols. There is a need to develop a set of common tools that will serve the various state, regional, and international registries in economic regulation and business decision-making. The goal of such tools is to realize a common currency for GHG reduction credits. The development of these tools must be accomplished through a fair and "registry-neutral" process that will not find favor in any single registry. Ultimately, it is important for the State and regional registries to achieve consensus around the development of common metrics.

Consistent economic tools for measuring and verifying emissions reductions among the various registries for GHG & RECs would facilitate project finance and investment and the development of environmentally sound projects. Tools that allow for consistent economic recognition of these credits would facilitate development of project "templates," thereby reducing costs and allowing rapid dissemination of the learning gained from early projects. Further, economic tools would make possible the growth of liquidity necessary to calculate future value of GHG reduction credits to projects, thereby supporting development and project financing.

To function efficiently, such tools would require assurance of integrity, clear definitions, avoidance of double counting, consistent verification methods, liquidity, and consistent treatment of metrics.

With coordinated, interconnected conversion methods and other economic tools among state/regional registries, it may be possible in the future to use carbon dioxide reduction credits recorded in a U.S. registry to meet the needs of Kyoto signatory nations or any other successors to the Kyoto Protocol. The European Union (EU) has taken steps in this direction with the implementation of an Emission Trading Scheme (ETS) that will integrate the Kyoto-treaty requirements with the EU's ETS to assure compliance with the treaty and at the same time, allow for uniform trading of GHG credits. The scheme, as announced, addresses many of the issues that must be taken into account in the establishment of a consistent set of rules and metrics for state registries (including double counting, limiting reduction and offset credits to "real" projects, eliminating "free riders," and facilitating the legitimate conversion of EU's credits to Kyoto credits, among others). Consistent U.S. registries would facilitate future trading not only with Europe, but also Canada and Japan.

As a result of these European and other U.S. efforts, there is now a large body of work to support the development of economic tools so that credits in State and regional registries can be used in conjunction with the EU ETS and other systems. Further, these tools will take on additional significance at such time that the U.S. may decide to take part in international mandatory efforts to reduce GHG.

Once economic tools are developed, state-mandated and/or regulated projects would be able to bank credits now in any registry for future use. If no internationally acceptable economic tools are developed, then these credits (even if required by state law) may never be recognized in the international community and would have a lesser financial value for U.S. companies. In any event, not taking the first step to establish a common currency and other economic tools in GHG reduction credits among the various state/regional registries would be a further disincentive for U.S. firms and international corporations operating in the U.S. to take action now.

BUILDING ON THE NEW ENTHUSIASM FOR RENEWABLE ENERGY

Renewable energy credits (RECs) are now going from promise to reality. The Texas and California REC markets have been extremely active. Today, 19 states have Renewable Portfolio Standards (RPS), and we are starting to see real market growth. As more states adopt renewable energy credit programs and renewable portfolio standards, many more trades will occur. Demand is also coming from commercial and industrial customers seeking green energy, with many active green power marketers stepping up to meet this market need. In addition, state governments are making purchases under renewable energy procurement mandates, with some federal agencies also participating.

With a new term for the Bush administration, it's time to set the framework to begin a national REC market and promote rapid commercialization of new technology. The nascent, homegrown U.S. renewable energy industry is presently in a state of confusion. The wind-power Production Tax Credit (PTC) went through an on-again-off-again period, wreaking havoc on the U.S. wind industry. While the PTC has now been extended through 2006, a long-term act is clearly needed to jumpstart the industry. Today, the U.S. small wind power business employs only 3,000 employees, compared to 45,000 workers in Germany where the rules are in place. Ironically, in the summer of 2004, there were shortages of renewable energy credits as consumers are finally embracing green power initiatives (with cost premiums) in a major way.

The U.S. government needs to set the rules for corporate America to invest in the renewable energy arena. Despite platitudes of national security and energy independence, all efforts to gain traction for renewables at the federal level have continued to be frustrated. Meanwhile, an interesting collection of diverse interest groups has assembled, working together to gain a toehold in the emerging renewable energy space. These groups range from renewable energy and energy efficiency policy wonks to agribusiness (in the form of the ethanol lobby) and defense contractors. This is because the renewable energy industry has reached a stage where there are technological breakthroughs every year. It has gone past the solar collectors and abandoned wind turbines of the early 1980s and produced the next generation of environmentally benign, cost-effective, and highly efficient renewable-energy technology.

The irony is that the current renewable energy industry is being funded by the Pentagon which has assumed the role NASA played in the 1970s. Many of the new technologies that are being readied for deployment have been funded by the Department of Defense and, to a lesser extent, the Department of Homeland Security. The threat of energy security is now real, and energy prices have been high and will remain so. Logically, the time for renewables and alternative energy is *now*.

INTEGRATING ENERGY EFFICIENCY INTO SUSTAINABLE ENERGY PLANNING AND ENVIRONMENTAL TRADING

Energy efficiency and demand-side management programs, while proven effective and much applauded, have generally been isolated from the mainstream of energy planning and use. However, there are hopeful signs of their integration into sustainable energy planning and programs to address climate change.

In the U.S., Pennsylvania became the first state with a clean energy portfolio standard that includes demand-side management, and some utilities have begun to seek power supply proposals that include demand-side resources. In California, a proposed rule would require the reporting of carbon dioxide emissions reduction associated with energy efficiency measures.

In Italy, Britain, and New South Wales (Australia), there are white or energy efficiency certificate trading schemes that are part of the overall GHG abatement efforts (see Chapter 6). Further, there is ongoing work to develop these approaches as well as explore their interactions/integration with other certificate trading schemes (e.g., green or renewable certificates) and the carbon/GHG markets. Such discussions and collaborations are taking place at the national and multinational levels (among European Union countries and through the International Energy Agency's Demand-side Management Programme).

MARKET OPPORTUNITIES AND RISKS IN THE TRANSITION TO A CARBON-CONSTRAINED ECONOMY

We are at the beginning of a major economic and social transition. For the financial markets and financial players, climate change and our response bring many new risks and new opportunities. On the negative side, new risks include carbon liabilities as well as potentially impaired credit quality of GHG-intensive borrowers. On the positive side, new or additional opportunities will open up for financing infrastructure development and clean technology as well as unprecedented opportunities to trade in the estimated $2 to $3 trillion GHG markets. Very importantly, these "Green Revenues" can enhance capital market liquidity as well as generate economic development.

Today, markets in renewable energy and GHG reductions are being created at the same time in different parts of the world. Mimicking the oil market developments of the late 1970s, the shift to a global market is evolving. Key to these markets is a recognized certification system with broad acceptance. Well-defined national renewable and GHG certification systems will facilitate trade and return maximum value to projects. Besides lowering cost, a common "currency" will expand project finance opportunities. In the U.S., we have begun to embark on this road, with 28 states working on GHG initiatives and nearly 20 states adopting or developing renewable portfolio standards, and the beginning of load management or demand response programs for trading in energy efficiency at some of the independent system operators.

Inevitably, such global development also engenders new competition and opens up unique opportunities for the right local players. For instance, based on the strengths of its financial services industry, New York City is in an exceptional position to emerge as the environmental finance center, leading the development of new Green Trading markets. Right here, in New York State and the Northeast, the Regional Greenhouse Gas Initiative is designing what would likely be the first U.S. mandatory cap-and-trade GHG market. At its doorstep, Wall Street has the extraordinary opportunity to facilitate the development and trading liquidity of this brand-new Green Trading market.

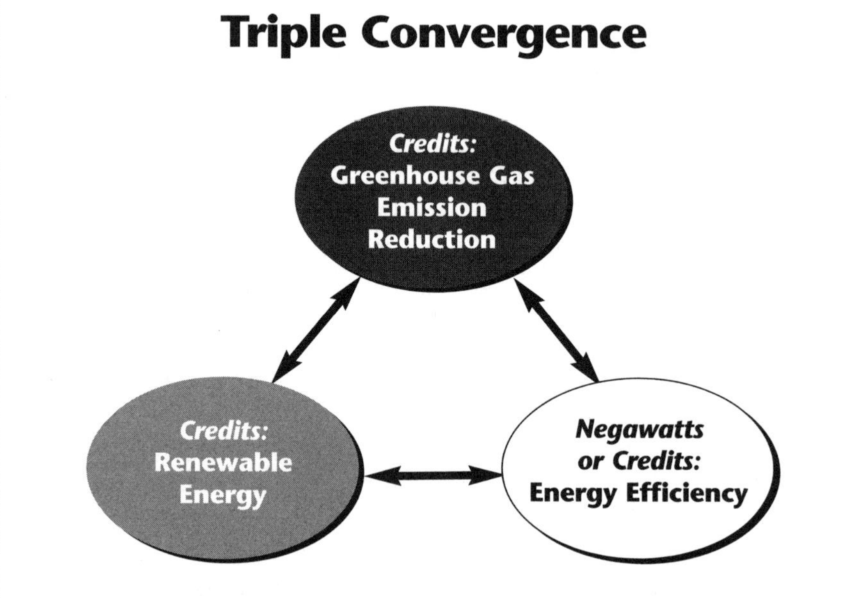

The Triple Convergence that we foresaw is now taking shape as the second wave of Green Trading takes off. Energy efficiency generates carbon reductions. Renewable energy reduces carbon emissions. On the balance sheet, carbon emissions footprints are being viewed alongside financial data.

2005 promises to be a year of global innovation and experimentation as the Green Trading markets' maturation process is finally accelerating. The technology exists to move forward, and the established financial players can learn quickly how to trade the new financial products with greater price transparency, market liquidity, and cost reductions. Clean, green energy is now in focus for institutional investors, energy hedge funds, and venture capitalists, and for the first time, it will be considered at the G8 Summit.

Peter C. Fusaro, *Founder* and *Chairman* of Global Change Associates, an international energy and environmental consulting firm, can be reached at T: 1 (212) 316-0223 and E: peterfusaro@global-change.com.

Marion Yuen is *President* of The MYA Group, a firm dedicated to creating and supporting partnerships that extend business competitiveness. She can be reached at T: 1 (718) 230-5402 and E: myuen@mya-group.com.

The authors co-created the **Annual GreenTrading Summit™: Emissions, Renewables & Negawatts** (www.greentradingsummit.com).

Resources on Green Trading

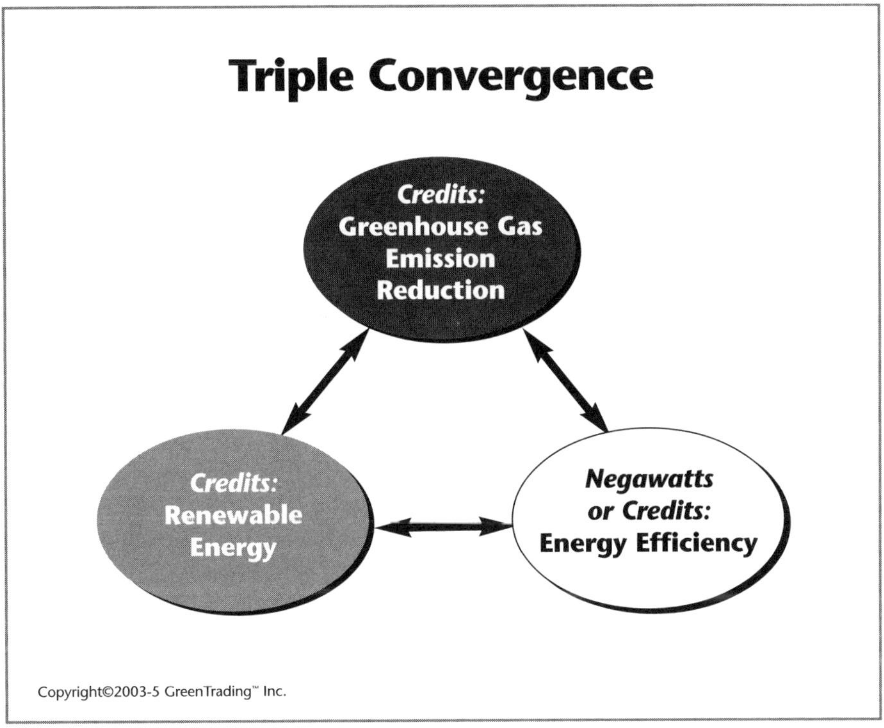

GLOSSARIES

Green Trading is a concept that embraces multiple areas. Explanation of basic ideas and terminology in the various fields is available from sources on the Internet, including the following:

California Energy Commission
http://www.energy.ca.gov/glossary/index.html

Cantor Fitzgerald Environmental Credit Trading Glossary
http://www.emissionstrading.com/glossary.htm

Evolution Markets
http://www.evomarkets.com/index.html

GreenFacts.org
http://www.greenfacts.org/studies/climate_change/toolboxes/glossary.htm

Intergovernmental Panel on Climate Change
http://www.ipcc.ch/pub/gloss.htm

Pew Center on Global Climate Change
http://www.pewclimate.org/global-warming-basics/full_glossary/terms_a.cfm

U.S. Environmental Protection Agency
http://www.epa.gov/airmarkets/trading/basics/#what
http://yosemite.epa.gov/oar/globalwarming.nsf/uniqueKeyLookup/SHSU5BUMCG/$file/glossary.pdf?OpenElement

ORGANIZATIONAL RESOURCES

The Green Trading community is diverse—comprising of private businesses and public organizations, established corporations and emerging players. The following is a selection of entities with websites that contain information relating to some aspect of green trading.

Trading-Related Organizations

Chicago Climate Exchange
http://www.chicagoclimateexchange.com

CO2e.com
http://www.co2e.com

Ecosecurities
http://www.ecosecurities.com

Emissions Marketing Association
http://www.emissions.org

Emissions Trading Handbook
http://www.etei.org

Environmental Resources Trust (ERT)
http://www.ert.net

European Union Greenhouse Gas Emission Trading Scheme (EU ETS)
http://europa.eu.int/comm/environment/climat/emission.htm

Evolution Markets
http://www.evomarkets.com

GreenTrading™
http://www.greentrading.biz

International Emissions Trading Association
http://www.ieta.org

Natsource
http://www.natsource.com

U.S. Environmental Protection Agency, Clean Air Markets
http://www.epa.gov/airmarkets/index.html
http://www.epa.gov/airmarkets/trading/index.html
http://www.epa.gov/airmarkets/capandtrade/index.html

U.S. Regional Organizations and Initiatives

California Climate Action Registry
http://www.climateregistry.org

Clean Air-Cool Planet
http://www.cleanair-coolplanet.org

(The) Climate Trust
http://www.climatetrust.org

NESCAUM (Northeast States for Coordinated Air Use Management)
http://www.nescaum.org

Regional Greenhouse Gas Initiative
(An Initiative of the Northeast & Mid-Atlantic States of the U.S.)
http://www.rggi.org

South Coast Air Quality Management District 's RECLAIM
(REgional CLean Air Incentives Market)
http://www.aqmd.gov/reclaim

West Coast Governors' Global Warming Initiative
http://ef.org/westcoastclimate

Texas Natural Resources Conservation Commission (TNRCC)—The Houston/Galveston Area (HGA) NOx Emission Allowance Program
http://www.tceq.state.tx.us/subject/subject_air.html

U.S. Government

Agriculture and Climate Change, U.S. Department of Agriculture
http://www.nrcs.usda.gov/technical/land/pubs/ib3text.html

Defense Environmental Network and Information Exchange (DENIX)
http://www.denix.osd.mil/denix/Public/Library/Climate/cseq.html

Carbon Sequestration Program, U.S. Department of Energy
http://fossil.energy.gov/programs/sequestration

Carbon Sequestration in Terrestrial Ecosystems (CSiTE), U.S. Department of Energy
http://csite.esd.ornl.gov/index.html

Clean Air Markets, U.S. Environmental Protection Agency
http://www.epa.gov/airmarkets/index.html

Climate Change Research Division, U.S. Department of Energy
http://www.science.doe.gov/ober/CCRD_top.html

Economic Research Service, U.S. Department of Agriculture
http://ers.usda.gov/briefing/globalclimate/index.htm

Energy Information Administration Energy Glossary, U.S. Department of Energy
http://www.eia.doe.gov/glossary/glossary_main_page.htm

Goddard Institute for Space Studies
http://www.giss.nasa.gov/

National Energy Technology Laboratory, Climate Change Policy Support, U.S. Department of Energy
http://www.netl.doe.gov/products/ccps/index.html

U.S. Government Climate Information and Services
http://www.climateservices.gov/

U.S. Environmental Protection Agency
http://yosemite.epa.gov/oar/globalwarming.nsf/content/index.html

U. S. Department of Energy (DOE)
http://www.energy.gov

White House Policy on "Ensuring Reliable, Affordable and Environmentally-Sound Energy"
http://www.whitehouse.gov/energy

More Resources

American Forests
http://www.americanforests.org/resources/ccc

CDMCapacity.org
http://www.cdmcapacity.org/index.htm

Clean Development Mechanism
(United Nations Framework Convention on Climate Change)
http://cdm.unfccc.int

ClimateBiz
http://www.climatebiz.com/

Climate Change Capital
http://www.climatechangecapital.co.uk

Climate Change Central
http://www.climatechangecentral.com

CO_2 Capture Project
http://www.co2captureproject.com

Ecosystem Marketplace
http://ecosystemmarketplace.net/index.php

European Commission—Environment
http://europa.eu.int/comm/environment/index_en.htm

Environmental and Energy Study Institute
http://www.eesi.org/index.html

Intergovernmental Panel on Climate Change
http://www.ipcc.ch

Forest Trends
http://www.forest-trends.org

International Institute for Sustainable Development
http://www.iisd.org

(The) Katoomba Group
http://www.katoombagroup.org

MIT Carbon Capture and Sequestration Technologies Program
http://sequestration.mit.edu

MIT Joint Program on the Science and Policy of Global Change
http://web.mit.edu/globalchange/www

Natural Resources Defense Council—Clean Air and Energy
http://www.nrdc.org/air/energy/default.asp

Pew Center on Global Climate Change
http://www.pewclimate.org

Prototype Carbon Fund
http://carbonfinance.org/pcf/router.cfm?Page=Home

Rocky Mountain Institute—Energy
http://www.rmi.org/sitepages/pid17.php

UNEP Climate Change
http://climatechange.unep.net

United Nations Framework Convention on Climate Change
http://unfccc.int

U.S. Global Change Research Information Office
http://www.gcrio.org

World Energy Council
http://www.worldenergy.org/wec-geis

World Resources Institute—Climate Change
http://climate.wri.org

World Wildlife Fund—Climate Change
http://www.worldwildlife.org/climate

Worldwatch Institute—Reducing the Threat of Climate Change in the U.S.: A Survey of Activities
http://www.worldwatch.org/features/climate/activities

About the Contributors

Peter C. Fusaro is the best-selling author of *What Went Wrong At Enron;* publisher of Fusaro Focus, a bimonthly financial newsletter; and chairman and founder of Global Change Associates, Inc., an energy and environmental advisory located in New York City since 1991. For 30 years, he has been at the forefront of energy and environmental change for 30 years. His accomplishments include working on the lead phase-down of gasoline in the 1970s; fuel reformulation in the early 1990s in the United States as well as in the Asia Pacific region; creating energy efficiency programs for the NY/NJ Port Authority, including the World Trade Center; pioneering gas and electricity energy efficiencies programs for Con Ed (Enlightened Energy) and Brooklyn Union; identifying fuel-switching opportunities to natural gas for utility and industrial customers; and, for the past 14 years, working on GHG reductions. He has been advisor to the U.S. State Department as well as the U.S. EPA and U.S.DOE, the Japanese government, the World Bank, and many energy companies around the world on the energy and environmental change.

Mr. Fusaro is co-editor and contributor to the book, *GreenTrading: Commercial Opportunities for the Environment* (2004). He has authored several books on energy risk management including *Energy Hedging in Asia* (MacMillan, 2005), *Energy & Emissions: Collision or Convergence* (John Wiley, 2005), *Energy Convergence* (John Wiley, 2002), *Energy Derivatives* (2000), and *Energy Risk Management* (McGraw-Hill, 1998), as well as co-authoring *What Went Wrong at Enron* (John Wiley, 2002). Previous employers include ABB Financial Services, Petróleos de Venezuela, the NYC Mayor's Energy & Telecommunications Office, and the U.S. Department of Energy. Mr. Fusaro has an M.A. in international relations from Tufts University and a B.A. from Carnegie-Mellon University.

Marion Yuen is president of The MYA Group which is a communications, research, and mediation advisory. A skilled facilitator and conference convenor, Ms. Yuen brings together diverse audiences and provides them with platforms for open and efficient face-to-face information exchange. Together with Mr. Fusaro, she created the Annual GreenTrading Summit™. Ms. Yuen also contributed to and edited the book *GreenTrading: Commercial Opportunities for the Environment.* She organized a coalition of major power generators that sponsored the first Executive Conference on Safety in Fossil Power Plants (2001). She produced the 1999 Global Fuel & Power Forum, 1998 Global Electric Power Forum, 2000 & 1999 Biotechnology Investing Conferences, and in 1991, Ms. Yuen co-chaired the Advertising Research Foundation's first Asian, Black, and Hispanic Research Workshop. Ms. Yuen holds master's degrees in Communications from Columbia University Teachers College and in Biophysics from the University of California at Berkeley, and an A.B. in Physics from Wellesley College as well as a certificate in Strategic Environmental Management from New York University's Management Institute. Ms. Yuen is a certified mediator and arbitrator.

Stefano Alaimo is head of the Environmental Markets Department at Gestore Mercato Elettrico (Rome), the Italian electricity market operator. His activities include development and management of an organized marketplace for Green Certificates and Energy Efficiency Certificates; analysis of the Emission Trading System in Europe and links with other environmental markets; analysis of international standardized and OTC derivatives markets on electricity, commodities, and weather; and preliminary analysis for the development of an exchange traded derivatives market on electricity prices in Italy. Before his involvement with GME, Mr. Alaimo was a proprietary trader on interest rates and FX markets at ABB Financial Services (Italian branch, Milan) where, along with pure trading activity, he has been involved in managing FX Forward book resulting from ABB FS activity with Italian subsidiaries. Previously, Mr. Alaimo worked as market maker on Italian government bonds at Smith Barney Europe Ltd. (London), with duties on long maturity bonds trading and basis trading. He also worked as fund manager at Fondicri Spa (Rome), a mutual fund managing company, and as a mathematics teacher at a high school. He graduated (magna cum laude) in Economic and Statistical Sciences at University "La Sapienza" (Rome).

David Brand is director of the New Forests Program with the Hancock Natural Resource Group, based in Australia. He is responsible for the design and oversight of forest investment programs that will bring together elements of traditional forestry investment with new environmental markets for forests including carbon sequestration, watershed management, and biodiversity enhancement. Prior to his current position, Dr. Brand was Executive General Manager of State Forests of New South Wales (NSW). In that position, he was active in supporting the development of carbon trading in NSW. From 1985 to 1995, he worked with the Canadian Forest Service as a scientist, as Director of Scientific Programs, and ultimately as National Director-General of Science and Sustainable Development. Dr. Brand has won several awards including the Canadian Forestry Scientific Achievement Award and the Canadian Public Service Award of Merit. He has authored a wide range of publications, books, and journal articles. He has a Ph.D. from the University of British Columbia and a B.S. from the University of Toronto.

Dawn Browne is the Geographic Information Systems manager for the national headquarters of Ducks Unlimited, Inc. (Memphis, Tennessee). She is responsible for coordination of national and international GIS and remote sensing projects as well as coordinating support for other GIS-related activities throughout the organization. Currently, Ms. Browne is working on a comprehensive carbon tracking system for Ducks Unlimited (DU). She has worked with DU for approximately five years as a GIS/remote sensing analyst in the southern regional office; currently she is a GIS manager. As an analyst, Ms. Browne developed several GIS data sets and models for the Lower Mississippi Alluvial Valley Joint Venture and the Gulf Coast Joint Venture, including Historical Forest Cover Change, a Soil Moisture Index, and a Redhead Freshwater Wetland Model for the Laguna Madre region. She has also been the technical coordinator for the multinational Upper Paraguay River Basin GIS project. Ms. Browne has a master's degree in Environmental Remote Sensing from University College London, England, and a B.S. in Environmental Science from University of North London, England.

Susanne Haefeli has worked exclusively since 2001 in the climate change field. She has worked on the development of accounting and reporting standards along the value chain and for CO_2 capture and storage projects. She also has managed the WBCSD's first phase of the CDM capacity building project, including the organization of a project developer's forum in South Africa. Ms. Haefeli presently is working in DNV Certification as a project manager on CDM and JI project validations and with other CDM and JI related work.

Paul A. Hilton is a portfolio manager for socially responsible investing at the Dreyfus Corporation. Mr. Hilton conducts social and environmental research for all current and prospective holdings in the Dreyfus Premier Third Century Fund and its variable annuity counterpart, the Dreyfus Socially Responsible Growth Fund. Mr. Hilton joined Dreyfus in August 1998. He left the firm in April 2001 and returned July 2002 to continue his management of the socially responsible portfolios. Prior to joining Dreyfus, Mr. Hilton was a research analyst in the Social Awareness Investment (SAI) program at Smith Barney Asset Management, a division of Citigroup. Mr. Hilton also served as project director for corporate social responsibility research at the Council for Economic Priorities (CEP), a New York-based not-for-profit organization best known for its consumer guide *Shopping for a Better World.* While at CEP, Mr. Hilton spent four months with the Centre for Science and Environment, an environmental research organization in New Delhi (India), developing a program to evaluate the environmental performance of India-based companies. Mr. Hilton holds a B.A. from Syracuse University, an M.A. in Anthropology from New York University, and an M.Ed. from Roberts Wesleyan College.

Sheila Slocum Hollis, a managing partner of the Washington, D.C., office of Duane Morris LLP, serves on the firm's partners' board. Ms. Hollis practices in the areas of energy transactional and regulatory law, and international and administrative law before government agencies, Congress, and other entities. She specializes in domestic and international energy, water, and environmental matters, representing governmental bodies and the power and natural gas industries. She recently was named one of the 50 Key Women in Energy.

Ms. Hollis is a board member of the United States Energy Association and a member of the Advisory Committee of the North American Energy Standards Board. A Fellow of the American Bar Association (ABA), she serves on the editorial board of the *ABA Journal* and on the Standing Committee on the Federal Judiciary. In addition, she served as chair of the Standing Committee on Environmental Law from 1997 to 2000 and served two terms as chair of the ABA's Coordinating Group on Energy Law. The first director of the Office of Enforcement of the Federal Energy Regulatory Commission, serving from 1977 to 1980, Ms. Hollis was a trial lawyer at the Federal Power Commission from 1974 to 1975. Over the course of her career, she has played a key role in the formation and implementation of energy law and policy. She is the immediate past chair of the Section of Environment, Energy and Resources of the ABA, representing 11,000 members. Ms. Hollis was the first woman to serve as president of the Federal

Energy Bar Association (1991–1992). Since 1997, she has served as president of the Women's Council on Energy and the Environment. Ms. Hollis served as a professorial lecturer in the law on the subject of energy law at the George Washington University Law School from 1979 to 1999.

Ms. Hollis is a member of the American Law Institute and serves on the Board of Trustees of the Center for American and International Law and as vice chair of its Institute for Energy Law. With an extensive background in international energy law, Ms. Hollis is a member of the Energy, International, Women's and Federal bar associations. Widely published in energy law and policy matters, having co-authored two energy law texts and numerous articles, she was ranked by the National Law Journal as one of the Nation's top 20 energy lawyers. A Colorado native, she is a graduate of the University of Denver College of Law and a graduate of the University of Colorado at Boulder.

Ed Holt is president of Ed Holt & Associates, Inc., a renewable energy policy and green power marketing consulting firm located in Harpswell, Maine. He has recent or current projects with NARUC, the U.S. EPA, AWEA, NRECA, EPRI, and NREL. He is also an active member of the National Wind Coordinating Committee. He has a particular interest in policy issues affecting the use of renewable energy certificates in voluntary and mandatory energy markets and in emission trading markets. Mr. Holt is the author of the *Green Pricing Resource Guide* and numerous reports on green power. In 1994–1995, he was a principal with the nonprofit Regulatory Assistance Project. For 12 years before that, Mr. Holt worked for Seattle City Light, planning and implementing customer energy-efficiency programs.

Claire Jahns is an associate economist at the Chicago Climate Exchange (CCX), where she works primarily in the marketing department. In her role, Ms. Jahns works with potential CCX members in the industrial and power generation sectors to interpret CCX rules and calculate their greenhouse gas emissions footprints, both of which are key elements to the recruiting and new business development process. In addition, Ms. Jahns serves as the primary point of contact for some CCX members and associate members during the annual True-up process and with trading and compliance issues. Along with her marketing responsibilities, she contributes to the research department, creates marketing materials and presentations, and works on industry analysis. Ms. Jahns holds a B.A. in economics and environmental studies from Oberlin College.

Dick Kempka is the director of Energy and Technology Partnerships for Ducks Unlimited, Inc. (DU). Based in Memphis, Tennessee, he is responsible for coordination of DU carbon sequestration projects and other eco-asset management activities throughout the organization. He has authored or co-authored more than 20 professional publications. Mr. Kempka has worked for approximately 10 years as a remote sensing analyst and, later, as GIS Director at the DU California office. As DU's GIS Director, he designed an entity-wide conservation database management system. Mr. Kempka spent many years developing GIS databases for DU's western regional office, including many projects/fieldwork in California and Alaska. He previously worked for Pacific Meridian Resources, a GIS consulting firm in Oakland, California, where he was remote sensing manager and led all wetland projects. Mr. Kempka also has worked for the Defense Mapping Agency (now NEMA) and the State of Idaho. Mr. Kempka has an M.S. in Geography and Remote Sensing from Indiana State University, Terre Haute, and a B.S. in Geography/Remote Sensing from Carroll College in Waukesha, Wisconsin.

Arthur Lee is principal advisor for Global Policy and Strategy in Corporate Health Environment Safety of the ChevronTexaco Corporation. Based in San Ramon, California, his responsibilities include the corporation-wide formulation of strategic positioning and policy development on issues ranging from ChevronTexaco's internal energy policy and air pollution issues to actions addressing climate change concerns. Mr. Lee is the leader of the Policy & Incentives Team of the CO2 Capture Project, which is the global joint industry-governments project to develop the next generation of CO2 capture and storage technology. Mr. Lee was the chairman of the Climate Change Working Group of the Regional Association of Oil and Natural Gas Companies in Latin America and the Caribbean from 1999 to 2003. He was the chairman of the American Petroleum Institute's Industrial Combustion Coordinated Rulemaking Task Force from 1996 to 1998, which responded to the series of rulemakings by the U.S. EPA on air toxins from primarily combustion sources. Prior to the merger of Chevron and Texaco, Mr. Lee was the manager of Strategic Business Initiatives and Environmental Analyses, responsible for developing and implementing programs in carbon sequestration, emissions reduction, and offsets generation, and related energy and environmental technology development for Texaco, Inc. He has held positions of increasing responsibility with Texaco's exploration and production group and in safety, health, and environment. Previously, Mr. Lee was a regulator at the U.S. EPA Acid Rain Division, where he participated in developing the proposals regulating nitrogen oxides emissions from utilities and participated in the design of the emissions trading and tracking systems. He holds an M.S. in Chemical Engineering from the California Institute of Technology and an S.B. in Chemical Engineering from the Massachusetts Institute of Technology.

Mark M. Little is Vice President of GE Energy's power generation segment headquartered in Schenectady, New York. Power generation had 2004 revenues of $6.1 billion. GE Energy is a world leading supplier of power generation equipment including gas, steam, wind and hydro turbine-generators, turnkey power plant services, gasification technologies and IGCC (integrated gasification combined cycle). In addition to Schenectady, key business locations include Greenville, South Carolina; Bangor, Maine; Tehachapi, California; Houston, Texas; Lachine, Canada; Salzbergen, Germany; Belfort and Bourogne, France.

Mark joined GE's Turbine Business in 1978. After holding several management positions in engineering, he was named Product General Manager for generators in 1989. In 1991, he became General Manager—Business Development for GE Energy, responsible for strategic planning and joint venture development. In 1992, he was appointed Product General Manager for gas turbines and in 1994 was named Vice President, Power Generation Engineering. He assumed responsibility for the large turbine generator segments of power generation in 1997 and in 2004 was named to lead the combined thermal, wind and hydro power generation group.

Mark holds BS and MS degrees in mechanical engineering from Tufts and Northeastern universities, respectively, and in 1982 earned a Ph.D. in mechanical engineering at Rensselaer Polytechnic Institute.

John Palmisano was one of a small set of reformers who developed the concept of emissions credit and emissions quota trading. From 1976 to 1982 he was an advisor to and a manager of the U.S. EPA, implementing market-based regulatory reforms. He has more than 25 years of experience in developing air pollution control programs for industry and regulators, developing emissions trading projects and policies, trading and brokering air credits, advising governments and industry on emissions trading, developing policies related to environmental auditing and environmental management, conducting environmental audits, air permitting, and developing air and water pollution control programs for industry and regulatory agencies. In 1984, Mr. Palmisano established AER*X, the first firm specializing in brokering emission credits. AER*X was started with funding from Wisconsin Power and Light and was eventually acquired by EDS and then, years later, by Enron.

For the last three years, Mr. Palmisano has been a partner in Energy & Communications Solutions LLC (Washington, D.C.), which is a project development company focused on energy efficiency, renewable energy projects, and environmental trading in Russia, Ukraine, and Central and Eastern European countries.

William G. Russell is president and CEO of SKN Worldwide-USA, Inc., a sustainability products and services company working with other organizations and nations to improve their economic, environmental, and social performance—their "triple bottom line." Mr. Russell applies combined environmental, corporate finance, technology, and management consulting experience to help clients increase productivity while consuming fewer material, energy, and financial resources. He previously worked for PricewaterhouseCoopers (PwC), where he designed, implemented, evaluated, audited, and provided assurance statements that focused on management systems, contingent liabilities, and financial reserves as well as disclosure statements covering environmental, financial, and sustainability reporting.

Mr. Russell led PwC's efforts to establish its global Environmental Management Information Systems and climate change-related practices, including the implementation of ERP, Document Management, Emissions Inventory, and Data Warehouse applications. After leaving PwC in 2000 to become the president of Ecos Technologies, he focused on the development of Internet-based environmental, health, and safety (EH&S) management and control systems for multinational companies. In 2002, he left Ecos to launch SKN Worldwide. He is a certified hazardous materials manager with over 20 years of experience. Mr. Russell has a B.S. in Chemical Engineering from the University of Maryland and an M.B.A. in Finance from Rutgers University. He is a member of the advisory board of the University of Michigan's Corporate Environmental Management Program (CEMP) and the technical advisory board of Innovest Strategic Value Advisors.

Richard L. Sandor is chairman and CEO of the Chicago Climate Exchange, a self-regulatory exchange that administers the world's first multinational and multisector marketplace for reducing and trading greenhouse gas emissions. Dr. Sandor is also a research professor at the Kellogg Graduate School of Management at Northwestern University. While on sabbatical from the University of California, Berkeley in the early 1970s he served as Vice President and Chief Economist of the Chicago Board of Trade. It was at that time Dr. Sandor earned the reputation as the principal architect of the interest-rate futures market. The City of Chicago and the Chicago Board of Trade honored Dr. Sandor for his contribution to the creation of financial futures and his universal recognition as the "father of financial futures." In August 2002, Dr. Sandor was chosen by *Time* magazine as one of its "Heroes for the Planet" for his work as the founder of the Chicago Climate Exchange. In November 2004, he was the recipient of an honorary degree of Doctor of Science, honoris causa, by the Swiss Federal Institute of Technology (ETH) of Zurich (Switzerland) for his pioneer work in the design and implementation of innovative and flexible market-based mechanisms to address environmental concerns. Dr. Sandor received his Bachelor of Arts degree from the City University of New York, Brooklyn College and holds a Ph.D. in Economics from the University of Minnesota.

Einar Telnes, M.Sc. and M.B.A., is technical director of DNV Certification's international climate change services. He has been active for several years with the early phases of the Kyoto mechanisms development and consulted to the UNFCCC as well as several governments and international organizations on monitoring, validation, verification, and accreditation issues related to GHG projects. Mr. Telnes has experience from nearly 100 AIJ, CDM, and JI project validations or verifications as well as verification of corporate GHG inventories for a number of large international companies. He was project manager for development of PCF's preliminary validation manual and the first issue of the ERUPT guidelines in the Netherlands; recently he has been responsible for development of the PCF/IETA validation and verification manual for JI and CDM projects. Mr. Telnes also has been principal on the development of the Australian Greenhouse Friendly programme and has taken part in the development of WBCSD's GHG protocols as well as the new ISO GHG guidelines.

Mark C. Trexler, Ph.D., has more than 22 years of energy and environmental experience. He has been part of the climate change field since 1988, when he joined the World Resources Institute in Washington, D.C. to work on the world's first carbon offset project. Dr. Trexler set up Trexler Climate + Energy Services in 1991 after moving to Portland, Oregon, where the firm is now based. For the last 13 years, the firm has been at the leading edge of innovative corporate GHG strategy development, GHG mitigation policy development, and the GHG project market. Working with companies ranging from

Japan's J-Power to Vermont's Stonyfield Farm Yogurt, Dr. Trexler's firm has helped dozens of companies position themselves for climate change. More recently, the firm has focused particularly heavily on helping companies and government anticipate the future GHG market and GHG credit prices.

Deltcho Vitchev is currently a director of Renaissance Finance International (RFI), a company that specializes in structuring and implementation of projects in environmental and energy fields. After a distinguished 15-year career as an engineer commissioning industrial installations in the power, chemical, and metallurgical industries, he started specializing in financial management of environmentally beneficial projects. He was appointed as advisor to the president of the European Bank for Reconstruction and Development. As a senior and then principal environmental specialist, Mr. Vitchev continued to work on the issues related to finance and environment. He continued his career as a principle banker of EBRD, financing projects in the field of energy and environment. Joining RFI, Mr. Vitchev continued his involvement in the areas of finance, energy, and environment by structuring and implementing projects in these areas and creating expertise in financial appraisal of such projects in Central and Eastern Europe. He is advising a number of international bodies on finance and environment, including the European Commission, the European Parliament, and the United Nations Economic Commission for Europe.

Patrick Zimmerman is director of the Institute of Atmospheric Sciences at the South Dakota School of Mines and Technology and is chair of the Department of Atmospheric Sciences. In 2002, Dr. Zimmerman helped create and was named interim director of the Center of Biocomplexity Studies, which represents a consortium of scientists from the University of South Dakota, South Dakota State University, and from South Dakota School of Mines and Technology. Dr. Zimmerman has been a principal investigator on over 24 major grants and contracts. His research has focused on biogenic hydrocarbon emissions and the carbon cycle. His patent application for method and apparatus for generating standardized carbon emission reduction credits was submitted on February 2003. From 1979 to 1997, he worked at the National Center for Atmospheric Research, where he helped to establish an interdisciplinary research program to study biosphere-atmosphere interactions. Dr. Zimmerman received his B.S. in Environmental Science (Zoology) and an M.S. in Environmental Science (Chemical Engineering) from Washington State University, and a Ph.D. from Colorado State University.

Index

D

E

F

G

J

K

L

M

N

O

T

U

V

W

About GreenTrading™ Inc.

GreenTrading™ Inc., was co-founded by Peter C. Fusaro and Marion Yuen after the successful 2002 GreenTrading™ Summit (www.GreenTradingSummit.com) in New York City.

Our mission is to provide information, analyses, and services that support the emergence and development of the GreenTrading™ business community.

GreenTrading™ will focus attention on the use of available and existing tools to achieve measurable environmental benefits and profit in both regulated and non-regulated contexts. We will:

- Promote market solutions that produce demonstrable benefits, thereby reducing energy and environmental uncertainty;
- Highlight positive results achieved by placing financial value on environmental actions;
- Lead the international dialogue and networking efforts to define and certify marketable products, including the development of standards and the identification of evaluation and certification bodies; and
- Support regulatory frameworks that recognize and encourage the contributions of market actions towards environmental improvement.

For more information, please visit www.greentrading.biz, or contact us at info@greentrading.biz or 1 (718) 230-5402.